Raising Sheep

Beginner's Guide to Raising Healthy and Happy Sheep

By: Irene Mills

Table of Contents

Introduction

Sheep are among the herd animals that people tend to like raising for many different reasons. Sheep can also be raised singularly or in small groups as pets. People enjoy sheep and sheep tend to gravitate toward people who they recognize. The author grew up with 500 head of sheep just to the south of their home. On days when the wind blew from the south, the family could smell there were sheep in the neighborhood. The memory of hearing the low *baas* of the sheep on summer days was peaceful and almost a quiet background noise for children playing everywhere.

Sheep have been domesticated for thousands of years for their wool, meat, and milk. Their wild counterparts come in shades of brown, while domestic sheep range in color from white to dark brown. Domestic sheep have shorter tails. All sheep have a distinctive *baa* to communicate within the herd and their immediate family.

Sheep are used for a wide variety of reasons, including milk, wool, lambing, and meat. There are many kinds of wool, including fine, long, medium, and carpet wool. Raising sheep for meat is more common in Europe than in the United States, but meat production is one of the uses in the United States. Sheep have been bred to have longer lambing seasons and for the production of more lambs. The industry is growing, but several breeds of sheep are on the endangered list.

Chapter 1: Kinds of Sheep

Kinds of Sheep

There are dozens of breeds of sheep. Several breeds are mixed with other breeds, then mixed with other breeds, and so on. In fact, the first mammal to be cloned was a sheep named Dolly. Many breeds go back as far as the records of time, and others have been developed as recently as in the last decade. Listed in this chapter are breeds that are among the most prolific and some that are the most endangered.

Many breeds are historic for the parts they might have played in the beginnings of a new continent. Sheep are historic for helping new continents begin and grow. When cared for correctly, they can live to be 4 to 8 years old.

<u>American Blackbelly</u> – The American Blackbelly is a low-maintenance breed that is well suited for the open range. This breed was developed in Texas by crossing Barbados Blackbelly sheep with Rambouillet sheep and Mouflon.

<u>Awassi</u> – The Awassi is the most abundant breed in southwest Asia. Today, it is the highest milk-producing breed in the Middle East. It is bred in desert conditions and lives in Saudi Arabia, Israel, Lebanon, and Jordan.

<u>Babydoll Southdown</u> – Babydoll Southdown sheep are raised primarily for pets. They are a smaller version of the Southdown breed. To be eligible for registry, sheep must be smaller than 24 inches (at the time of shearing). Babydolls are sometimes used for grazing in vineyards and in private yards.

<u>Barbados Blackbelly</u> – The first Blackbellies were introduced to the United States in 1904. The Barbados Blackbelly originated on the Caribbean island of Barbados. It descends from sheep brought to the islands from West Africa. Blackbellies are brown, tan, or yellow in color, with black markings. Both ewes and rams are polled or have only small scurs or diminutive horns. They are noted for their hardiness, including resistance to internal parasites, and reproductive efficiency.

<u>Black Welsh Mountain</u> – This breed was introduced to the United States in 1973. Semen imports in the late 1990s added genetic diversity to US flocks. This is a

small sheep that produces a dense, completely black fleece. Ewes are polled, but rams have beautiful horns.

Bluefaced Leicester – This breed was introduced by England to the United States during the early 1980s. Their wool is very fine against their deep blue skin. They produce long, fine wool.

Booroola Merino – The original Merino came from Spain, but this species has evolved. This is a type of Merino sheep that has multiple births. They were developed in Australia from a flock that was being selected for improved reproductive rate.

Border Cheviot – The Border Cheviot originated in the Cheviot Hills, on the border between England and Scotland. They were imported into the United States in 1838 when many people immigrated to the US. Cheviots are a small breed, with a white face and bare head and legs. They have erect ears, with a stylish and alert appearance. Cheviots are known for their hardiness and spirit. They produce medium wool.

Border Leicester – The Border Leicester is a dual-purpose, long-wooled sheep that originated in England. They are named so because they are near the border between England and Scotland. It is not known when the Border Leicester was first imported into the US, but it is believed to be before 1920. Their long, curly, thick wool is loved by hand spinners.

California Red – California Reds originated in California in the 1970s. Lambs are born red. Their fleece color lightens as they get older. Fleeces of mature animals are beige or oatmeal colored, with dark hairs interspersed. Because of the hairs in the wool, the wool isn't viable commercially; however, it is popular with hand spinners. Rams are polled, but usually have manes. The breed has an extended breeding season.

California Variegated Mutant – The Romeldale is a cross between the Romney and Rambouillet. The California Variegated Mutant (CVM) is a color pattern of Romeldale. The soft wool and the unusual colors of the CVM are especially valued by hand spinners. Romeldales and CVMs are generally considered two types of the same breed. While Romeldales are mostly white, CVMs are naturally colored.

Charollais – Charollais sheep originated in France in the same region as Charollais cattle. The United States has a small number of this breed of sheep. There are larger numbers in Canada.

Clun Forest – This sheep came to the US in the 1970s. The Clun Forest is one of the purest breeds of sheep in the UK. Its parent breeds are Hill Radnor, Shropshire, and Kerry Hill.

Columbia – The Columbia sheep were developed in the early 1900s by the United States Department of Agriculture, with the intent of replacing crossbreeding on the range. While originally developed for range conditions, the Columbia has proved adaptable to conditions throughout the US. They are one of the largest sheep breeds in the US. Columbia sheep are a dual-purpose breed that produce fast-growing lambs and heavy, medium wool fleeces with good staple length.

Coopworth – The first Coopworths were imported into the United States in the 1970s, and they are raised mostly for their wool. Coopworths are a medium size, dual purpose, long-wool sheep. Their wool is relatively coarse, with a long staple length. The American Coopworth Registry is the only registry that offers breeders a performance designation for their sheep.

Cormo – This breed was introduced to the US in 1976 where it is raised mostly for wool. The breed is fairly rare and you won't see the Cormo on most farms. The breed was developed in the early 1960s in Australia. The wool is very fine.

Corriedale – The Corriedale is one of the most numerous breeds worldwide. It is a dual purpose sheep, suitable for both lamb and wool production. This breed is best known in Australia and New Zealand but is also prevalent in the United States. Corriedales were first brought to the United States in 1914.

Photo by Wayne Jenkins[i]
Featured on corriedale.org

Cotswold – The breed was introduced to the United States in 1831. In the US, it is raised primarily for wool. The Cotswold originated in the Cotswold Hills in England. Cotswold wool hangs in a lustrous, silky sheen. There is a separate breed registry for Black Cotswold as they are considered a separate breed. The Livestock Conservancy categorizes the Cotswold as "threatened."

Debouillet – The Debouillet was developed in New Mexico in 1920 from Delaine Merino and Rambouillet crosses. The breed is well adapted to the range conditions of the southwestern United States. The Debouillet is raised primarily for its fine wool. It is still common to cross Merinos with Rambouillets. This sheep produces a fine wool.

Delaine Merino – These sheep were introduced into the United States in the early 1800s. This is one of the oldest breeds of sheep in the world and it is one of the most influential. The oldest Merinos were developed in Spain during the 12th century. They have an extended breeding season and their wool is popular. They are recognized by their curly ears.

Ram   Ewe

UK College of Agriculture[ii]

<u>Dorper</u> – Dorpers were imported into the United States in the mid-1990s. They originated from South Africa and they lack the parasite resistance of other US hair breeds. They also do not shed as well, sometimes leaving bunches of wool or retaining wool on their back. However, they are the heaviest muscled hair breed in the US and one of the most popular.

<u>Dorset</u> – The Polled Dorset is one of the most popular breeds in the US, but the Horned Dorset is not. Dorsets are an all-white sheep, medium in size. They produce medium wool. Dorsets, especially the horned variety, are best known for their ability to breed out-of-season.

Cheeg Prep Flashcards[iii]

East Friesian – Crossbred Friesians were introduced to the US in 1993; the first purebreds arrived in 1994. Friesians are raised as either purebreds or crossbreds on US dairy sheep operations. Purebreds are introduced into a flock to improve milk production. The crossbreds are much stronger physically than the purebreds. Lambs will sometimes catch pneumonia. They have medium wool.

Finnsheep – Finnsheep or Finnish Landrace, as they are known in their native country of Finland, are one of the most well-known breeds of sheep in the world. Typical litter sizes are three to four lambs. Finns are related to other Scandinavian short-tailed breeds. They were first imported into the United States in 1968, where their primary use was to produce crossbred ewes for commercial lamb production. This is a perfect sheep for a small family farm

Triple Finn Sheep[iv]

Florida Cracker – The Florida Cracker is considered to be one of the oldest breeds of sheep in the United States, having descended from the sheep Spanish explorers brought to Florida in the 1500s. The breed evolved under natural conditions, roaming free until the end of WWII, when Florida's open range policy ended. Because Florida Crackers are an unimproved breed, they vary in size and appearance. Many have red markings, similar to the Tunis. They produce a medium wool. They are categorized as a "critical" breed by the Livestock Conservancy.

Florida Cracker Sheep[v]

Gotland – The Gotland takes its name from the Swedish Island of Gotland where it originated. They are a medium-sized, naturally polled breed of the Northern European short-tailed variety of sheep. Colors range from silver to almost black. The Gotland has continued to be established in the United States since 2003 by importing semen.

Gulf Coast Native – The Gulf Coast Native is believed to have descended from the sheep brought to the Americas by Spanish explorers. They developed largely through natural selection under the humid, sub-tropical conditions common to the Gulf states. Numerous universities have documented the parasite resistance of the

Gulf Coast Native. Most of the sheep are white. They lack wool on their faces, legs, and sometimes bellies. They have medium wool.

Hampshire – Hampshire sheep were developed in Southern England by the mingling of various breeds, including Old Hampshire. These sheep were first imported into the United States in the 1860s during the Civil War, so many of the flocks were scattered or destroyed. Importations resumed in the 1880s. They are known for production of meat and have medium wool.

Herdwick – The Herdwick is an old breed, possibly originating from Scandinavia. They are considered to be the hardiest of Britain's hill sheep. Herdwick lambs are born mostly black. Their fleeces turn to dark brown and gray as they get older. Herdwick rams sport horns, while ewes are polled. They are a medium sized, dual purpose breed. In 2008, the first Herdwick semen was introduced to the United States. Their wool is thick, coarse, and wiry, used primarily for outer wear and carpets.

Hog Island – In the 1700s, a flock of sheep that originated from British breeds at the time was established on Virginia's barrier islands. For centuries, the sheep adapted to the island environment, free from human intervention, becoming feral. In 1978, the last sheep were removed from Hog Island. The sheep are being preserved by several organizations due to their relevance to American history. One flock is at Mt. Vernon, the past home of George Washington. Hog Island sheep vary in physical appearance. As with other landrace breeds, they are relatively small. Most are white while about twenty percent are black. Ewes and rams can be polled or horned. The Hog Island breed is classified as "critical" by the Livestock Conservancy. They have medium wool.

Icelandic – The Icelandic is one of the world's oldest and purest breeds of sheep. It is the only breed of sheep raised in Iceland. It is of the Northern European short-tailed group of sheep, having a short tail that is not docked. Rams and ewes may be horned or polled. Icelandic sheep are used for meat, wool, and milk. Their fleece is dual-coated and comes in a range of colors. Icelandic sheep were first imported into the United States in 1993.

Ile de France – The Ile de France is a breed native to France. It was developed at a French Veterinary College in the 1830s by crossing the English Leicester and Rambouillet. The Ile de France is a major breed in France and is popular worldwide. It is used primarily as a terminal sire. The United States also uses this breed for lamb production. This sheep produces medium wool.

Jacob – The Jacob sheep were introduced into the United States in the early 1900s. The American Jacob is small, like the British Jacob. The Jacob has four horns; two center horns, and two side horns curling alongside the head. The coloring is black and white fleece. Jacobs are categorized as "threatened" by the Livestock Conservancy.

Karakul – Karakuls were first introduced to the United States during the early 1900s. The Karakul is probably one of the world's oldest breeds of domesticated sheep. It is native to Central Asia. Karakuls differ radically in conformation as compared to most US breeds. Most lambs are born coal black, with lustrous wavy curls. The pelts of Karakul lambs were historically referred to as Persian lambskin. Most adult Karakuls have a double-coat. Karakul ewes have an extended breeding season. They produce a carpet wool.

Katahdin – The Katahdin is an American breed of hair sheep developed in the 1950s on the Piel farm in Maine. It is the result of crossing hair sheep from the Caribbean with various British breeds, namely the Suffolk. In the 1970s, the Wiltshire Horn was introduced to add size and carcass quality to the mix, but eventually selection was against using the horns. The Katahdin is one of the best all-around hair sheep in North America. Katahdin sheep excel in maternal and fitness traits, including parasite resistance. They are the most numerous registered breed of sheep in the United States.

Katahdin Sheep[vi]

Kerry Hill – Kerry Hill is a British breed that originated in a region near the Welsh/English border. It has unique markings, with a white face and black markings around its mouth, ears, and eyes. Its legs are also white with black markings. Some say this is a beautiful sheep. Kerry Hill semen was imported into the United States in 2006. Kerry Hill produce medium wool.

Lacaune – The Lacaune is the most popular dairy sheep in France. From its milk, France makes the Roquefort cheese. As compared to the East Friesian, the Lacaune produces less milk, but with higher total solids. Though the Lacaune is a wooled breed, it tends to shed most of its wool from the chest down. The Lacaune was imported into Canada in 1996. In the United States, it has become common to cross the Lacaune with the East Friesian for dairying.

Leicester Longwool – The Leicester Longwool (or English Leicester) is a longtime old English breed that produces a fleece that is heavy, curly, soft handling, and lustrous, with a spiral-tipped staple. The breed was developed in the 1700s by Robert Bakewell. The Leicester Longwool is a heritage breed, well-known in history by the American colonies. Conservation efforts are spearheaded by the

Colonial Williamsburg Foundation which maintains its own flock. The Livestock Conservancy categorizes the Leicester Longwool as "threatened."

Lincoln – The Lincoln is also an English breed and is considered to be one of the world's largest breeds of sheep. Its fleece is the heaviest, longest-stapled, and most lustrous in the world. Lincolns were first brought to the United States in the 1800s. They contributed to the development of several commercially-important American breeds including the Columbia and Targhee. In the United States, Lincolns are raised mostly for their long wool.

Montadale – The Montadale is an American breed of sheep developed by E.H. Mattingly in the 1930s. Mattingly's goal was to combine the best characteristics of Midwestern mutton-type sheep and big Western range sheep. He bred Cheviot rams to Columbia ewes. The Montadale is a large whiteface sheep. It is suitable as either a sire or dam breed. This sheep produces a medium wool.

Chegg Prep Flashcards[vii]

Navajo-Churro – The Navajo-Churro is one of the oldest sheep breeds in the United States. This breed of sheep descended from the Churra sheep brought to the New World by Spanish Explorers. Over the years, there were numerous attempts to destroy the Navajo-Churro population, almost resulting in the breed's extinction. Since the 1970s, the Navajo-Churro Project has been dedicated to preserving this culturally important breed.

Navajo-Churros are a hardy breed and adapted to the adverse conditions found in hot, dry deserts and sub-zero climates. They have a fleece that is classified as carpet wool and used primarily for rug weaving. The wool is the basis of the well-known Navajo carpets. Ewes and rams can be polled or have two to four horns. Despite continuing conservation efforts, the Navajo-Churro is still classified as "threatened" by the Livestock Conservancy.

Puddleduck Farm[viii]

North Country Cheviot – The North Country Cheviot is native to Scotland. They are larger than their southern relatives, the Border Cheviot. North Countries are a hardy hill breed that is raised primarily for meat. The first North Country Cheviots were imported into North America in 1944, where they were favored by shepherds looking for sheep that can take care of themselves. They produce a medium wool.

Oxford – The Oxford originated in England. It is the result of crossing Cotswolds with Hampshires. A small amount of Southdown blood was introduced in the early development of the breed. The first Oxfords were imported to North America in 1846. This is a medium to large sheep, with a dark brown face and wool on the legs. It is used primarily as a terminal sire. It produces a medium grade wool.

Painted Desert – The Painted Desert is a spotted hair sheep. It originated in Texas by crossing Mouflon with Rambouillet, Merino, and Texas Blackbelly. Some Jacob and Navajo Churro influence was used to create multi-horned animals. Painted Desert sheep breed out-of-season. They may be flighty when confined to small spaces. The breed is raised mostly for trophy hunting.

Panama – Panama sheep originated in 1912 in Idaho, United States. The foundation of the breed was a cross of Rambouillet rams to Lincoln ewes. Panamas produce a medium, long-stapled fleece. They are a hardy sheep, best adapted to range conditions. It is not known if many purebred Panama sheep remain. They produce a medium wool.

Perendale – Perendales were developed in the 1950s at Massey University in New Zealand. Their parent breeds are Border Cheviot and Romney. They were developed as an easy-care breed. They are considered dual-purpose and well-suited to cold, high rainfall areas. They produce a long wool.

Polypay – The Polypay was developed in the 1970s at the U.S. Sheep Experiment Station in Dubois, Idaho, and Nicholas Farms at Sonoma, California. Targhee x Dorset and Rambouillet x Finnsheep crosses were mated to form a 4-breed composite that could produce two lamb crops and one wool crop per year. Polypays are a medium-sized, prolific breed with an extended breeding season. Good mothers and milkers, they produce lambs with good growth and carcass quality. The name Polypay comes from "poly" for many or much and "pay" to indicate a return on investment and labor. The Polypay produces medium wool.

Racka – The Racka is a Hungarian breed of sheep known for its spiral-shaped horns. Mature males may have horns as long as 2 feet or more; the minimum standard is 20 inches for rams and 12 to 15 inches for ewes. The corkscrew horns protrude almost straight upward from the top of the head. The Racka has double-coated fleece, with a very long and coarse outer coat. Color varies from white to brown to black. The Racka has long wool.

Rambouillet – The Rambouillet is the foundation of most western range flocks. It was developed from the Spanish Merino in France and Germany and imported into the United States in the 1800s. The Rambouillet is a large, rugged breed, of medium growth, and good longevity. This sheep flocks well. This is an important trait in range flocks. Rambouillets produce high quality, fine wool. Though considered a range breed, the Rambouillet is adaptable to different conditions and is raised throughout the United States. They are considered to be a dual-purpose breed. The Rambouillet ewes have an extended breeding season.

Rideau Arcott – The Rideau Arcott is a Canadian breed. This breed is a mix of several breeds of sheep. The Rideau Arcott excels in reproductive traits: early puberty, prolificacy, and an extended breeding season. The breed is widely used in the Canadian sheep industry and was imported into the United States in the 1990s. The Rideau Arcott produces a medium wool.

Romanov – The Romanov is one of the most common breeds of sheep in the world. It originated in Russia's Volga Valley. The breed is of the Northern European short-tail variety. Purebred Romanov lambs are born pure black and lighten to a soft, silver gray as they make their fleece. The fleece of adult animals is double-coated and not well suited to commercial markets. Romanovs possess outstanding maternal characteristics. Such ewes will produce lamb crops between 200 and 300 percent.

Romney – The Romney is a long wool sheep that developed in a marshy area of England. It is the predominant breed in New Zealand. Romneys were first imported into the United States in 1904. The Romneys are raised mostly for wool in the United States. Their fleeces are heavy, long, and lustrous. Their wool has the finest diameter of any of the long wool breeds. Romneys can be white or colored. They are open-faced and polled.

Royal White – This sheep is an American breed of hair sheep. It was originally called the Dorpcroix. It was developed in the 1990s by William Hoag. It is a cross between the Dorper and St. Croix. The breed is pure white and grows a longer hair in the winter that is shed off naturally in the spring. Ewes and rams are naturally hornless.

Santa Cruz – The Santa Cruz once existed as a feral sheep on the Santa Cruz Island of the Channel Islands of California. After they were removed from the island, conservation efforts were begun to save the breed from extinction. Santa Cruz are a small, hardy sheep; most probably came from the Merino, Rambouillet, and Churra breeds. They are few in number.

Scottish Blackface – Numerically and economically, the Scottish Blackface is one of the most important breeds in the United Kingdom. In the United States, they are a minor breed, raised mostly for their wool. Scottish Blackfaces are horned in both sexes and as their name suggests, they have a black face, sometimes with white markings. They produce a carpet quality wool. They are a tough and adaptable breed, one of the most stunning in looks.

Shetland – The Shetland is a very old breed, having originated over 1000 years ago. It belongs to the Northern European short-tailed group of sheep. Just like the Shetland pony is small, so is the Shetland sheep. Shetlands are known primarily for their production of colorful wool upon which the Shetland woolen industry is based. Shetland wool comes in one of the widest ranges of colors of any breed. There are 11 main colors as well as 30 markings, many still bearing their Shetland dialect names. As a primitive breed, Shetlands naturally shed their wool during late spring/early summer.

Shropshire – The Shropshire breed originated in central western England. They have fine textured wool that is used in woolen hosiery and knitting yarns. The wool is used as spinning for hand knitting. The wool is also exported to Japan to fill futons. The Shropshire sheep is used for meat production.

Soay – These sheep have been in the United Kingdom since before the Roman Era. Soay sheep descend from feral sheep on the island of Soay in the St. Kilda Archipelago off the northern coast of Scotland. Soay means "sheep island" in Norse. They are small and quick. Rams and ewes can both be horned and ewes can also be polled. They shed their fleeces naturally. This breed does not flock well and they also do not do well with sheep dogs, which tend to scatter the sheep rather than group them.

South African Meat Merino – The South African Meat Merino originated in South Africa. It is derived entirely from the German Mutton Merino. Years of selection resulted in a sheep with good meat and wool. They produce a fine wool. This breed is now found throughout the world.

Southdown – The Southdown was developed in the Sussex hills of England in the late 1700s and early 1800s. They are known to be a healthy and hardy breed. The color of their face and legs is gray to brown. They produce a medium fleece that is the finest of the British breeds. They are one of the most popular breeds of sheep in the UK and the United States. They are a multipurpose sheep and produce milk, meat, and wool.

The Livestock Conservancy[xi]

St. Augustine – The St. Augustine is a relatively new breed of a hair sheep. This breed was developed in the 1990s by Ron & Ruth Tabor. It is a cross between the St. Croix and the Dorper. The St. Croix was selected because of its mothering ability, parasite resistance, and hardiness. This breed is well-adapted to hot, humid conditions and has good parasite resistance, perfect for Florida's climate. They have good flocking instinct and also have a nice disposition. The sheep are often spotted.

Livestock of the World[xii]

St. Croix – The St. Croix is a hair sheep native of the US Virgin Islands and named after the island of St. Croix. The St. Croix descends from the African sheep brought to the Caribbean on slave ships. Most of the breed is white with some solid tan, brown, black, or white with brown or black spots. Both ewes and rams are polled. In addition to having outstanding reproductive capabilities, the breed flocks well and is calm. The St. Croix breed is a good addition for a hobby farm.

Suffolk – The Suffolk is a British breed of sheep. This sheep breed is the result of crossing Southdown rams onto Norfolk ewes. This breed is common all over the world. Today, the Suffolk is one of the most popular breeds in the United States. It is the most popular sire of market lambs and one of the preferred breeds for producing club lambs. They have a bare black head and legs and produce medium wool.

Countryfile Magazine
"Native British Sheep Breeds and How to Recognize Them"[xiii]

Targhee – The Targhee is an American breed of sheep, developed in the 1920s at the U.S. Sheep Experiment Station in Dubois, Idaho. The Targhee is a cross breed of several hardy breeds. The Targhee derives its name from the Targhee National Forest on which the experiment station's flock grazed in the summer. The Targhee produces meat and a heavy fleece of high quality wool. The breed is especially popular in Montana, Wyoming, and South Dakota.

Teeswater – The Teeswater is a large, long-wool sheep from Great Britain. They are known for their long hair with a staple length of 8 to 12 inches. The wool is among the finest of the long wool breeds. It has a luster that remains after shearing and continues after spinning.

Texel – Texel sheep originated on the Isle of Texel off the coast of the Netherlands early in the 19th century. The Texel produces medium wool. It may be more resistant to internal parasites as compared to other wooled breeds, making it a good choice for siring crossbred lambs that will be grazing on grass. The Texel has become the dominant terminal-sire breed in Europe and is becoming increasingly popular for this purpose in the United States. Some people say certain Texels have a face resembling a dog.

Tunis – These sheep are the result of combining the Middle-Eastern fat-tailed sheep imported from Tunisia with the sheep brought to America around 1799 from the ruler of Tunis as a gift. The Tunis is a medium sized sheep with a creamy colored wool. Their head and legs are solid tan to cinnamon red in color. Lambs are born red and lighten as they mature.

<u>Valais Blacknose</u> – People believe this is the world's cutest sheep! This sheep is from Switzerland. They are used for both meat and wool. One can hear the bells around their necks on the mountain slopes because these hardy sheep still are in flocks on mountainous slopes in Switzerland and surrounding areas. They produce a long wool used in carpet production.

<u>Wensleydale</u> – The Wensleydale is a long-wooled sheep that originated in England. It is a large sheep with long-stapled, excellent wool that falls in ringlets almost to ground level in unshorn sheep. The breed also has a distinctive gray-black face, ears, and legs. This is one of the most popular wools to work with today. People who are spinners and weavers love the Wensleydale wool and the many colors they use to dye it and make it bright for their designs.

<u>Wiltipoll</u> – The Wiltipoll is a long-hair sheep that was developed mostly from the Wiltshire Horn, with an infusion of various wooled breeds. Primarily, they are a polled version of the Wiltshire Horn, developed in Australia as an easy-care alternative to wooled sheep. They are a large sheep breed used for their meat. They are hardy and excellent foragers.

<u>Wiltshire Horn</u> – The Wiltshire Horn is an old English breed that naturally sheds its short wool and white hair coat. This breed is popular with small farmers. As the name suggests, both ewes and rams are horned. The Wiltshire Horn's presence is documented to be in the United States at the time of the beginning of the formation of the country, during the 1600s. The breed was again imported to the US in the 1970s, when it was used in the development of the Katahdin breed. Importations in the 1990s freshened the genetic base

Wiltshire Horn Sheep Stud[xiv]

Chapter 2: Why Do You Want to Raise Sheep?

There are any number of reasons to raise sheep. This may be the most profitable animal to raise with the least amount of direct impact to planet Earth. As you found in the previous section, there are plenty of breeds to choose from for whichever reason you choose to own sheep.

Sheep as Pets

A sheep's lifespan is five to ten years of age depending on the care they receive. Sheep love to be in a flock, so in order to have happy pet sheep, you should have at least three or more so they can keep each other company. There is a story of one mom who let her sheep sleep inside until the animal was about six months old. When the sheep was older and bigger, she tried to move it outside to a stall to sleep at night. This completely backfired because the sheep believed itself to be human, and that it was supposed to sleep inside with its family, its herd.

The family received advice from a veterinarian and the advice was to get two to five more sheep and keep them all outside together. The problem was eventually solved and the separation anxiety was calmed down. Sheep make wonderful pets, but do not give them the idea that they are human or belong indoors. Sheep are outside, grazing animals. They have a wool coat for a reason.

Keep in mind that sheep are individuals similar to dogs and people. Some sheep are more independent than others and some are more aggressive than others. If you have young children, you should get lambs your kids can raise from up to adult sheep. Sheep will recognize people they see often and can learn their name. If you choose a ram, you should neuter the animal or it could become aggressive. However, most sheep are sweet and docile with gentile personalities.

Shelter for Pet Sheep

Build a small shed or barn with three sides for the sheep you have in your care. In the hotter months, you should provide shade from the sun. If the ewes lamb during the winter you will need a 4-sided shed.

Article with Instructions to Build
"A DIY Mobile Sheep Shelter"
Countryside Magazine Blog[xv]

Feed for Pets

Sheep will graze on grass and clover in fields and yards. If you overfeed with grain, they will get overweight quickly. Sheep are grazing animals. In the winter, they do just fine on hay. If your sheep are losing weight, then you might consider giving them grain.

Fencing

Fencing for a few pet sheep should not be too difficult. Protection from predators is what might make a fence necessary. Generally, pet sheep do not wander off, but choose to stay close to their area of grazing and stall.

Shearing

You have to shear your sheep every year in the spring before it gets too hot. This is something you cannot skip. Either find someone to do this task or take it on yourself, but there is no exception here. They have to be sheared. All the wool has to be shorn and the minute you get the sheep, you should make an appointment with someone to shear them. Everything around the reproductive parts has to be shorn because if you miss these areas, when it is hot, the sheep will be miserable and they will let you know.

Sheep for 4H

Young people who want to raise sheep for more than pets may also be members of 4H in their area. Sheep 4H projects provide youth with opportunities to learn more about the industry, production, and even more importantly, develop life skills to be positive members of society. These youth members learn about the care and raising of sheep in the most ecologically sound ways through nutrition, feeding, health, and daily responsibility.

4H kids learn hard work and the value of the care of animals. They learn to think critically, make decisions about exhibitions, and develop new friendships based on common interests. There are several exhibits to show sheep and the participants must learn good communication skills to talk with people who might ask questions at the show. Students in 4H make connections with older adults who are mentors and help them gain exposure to college and career opportunities.

Sheep for Meat

When you raise sheep for meat you need to go straight to the breeder to buy lambs. The breeder should tell you which lambs are bottle-fed and which are fed from their mothers. Buy the lambs that are fed from their mothers. These sheep will be healthier and weigh more as they grow.

You need to decide how much meat you want to produce. On average, the lamb will yield 40% of its live body weight in meat that is usable. You have to determine how many lambs you will need to raise to be able to produce what you need. A smaller the breed equals more work for you. Lambs are raised until they are about eight months old, then they are killed as humanely as possible and the meat is packaged for sale.

Sheep for Wool

Most farmers or shepherds who raise sheep for wool do not think of money when they first get into the wool/sheep business. People have been raising sheep for their wool since there have been sheep and humans on this earth. Wool is a renewable resource and there is something about raising an animal and shearing the animal for your good and for theirs that is gratifying.

Some shepherds also make clothes. The Navajo make their own woolen blankets from their Navajo Sheep. Learn first about the entire process before you make up your mind about raising sheep for wool. The rewards are not necessarily monetary. Depending on the type of sheep you raise, your location, and the amount of sheep you have, the wool can be worth a lot of money or not worth much at all unless you have a predetermined buyer.

Four Interesting Facts about Wool

- The crimp (waviness) of the wool is what makes it easy to work with, but it's also what gives wool clothing its bulk. Wool from sheep with a fine crimp tends to produce less bulky clothing, but it is more challenging to spin. The bulkiness of wool clothing is one reason why it provides excellent insulation.
- Wool can absorb almost one-third of its own weight in water but it retains its insulating properties even when wet.
- It takes a much higher temperature to ignite wool than it does to ignite cotton or synthetic fabrics and when wool does catch fire, the flame spreads very slowly. In fact, wool doesn't melt or drip in the presence of heat, it forms a "char." Men and women who are likely to encounter dangerous flames have traditionally worn wool clothing.
- Wool is hypoallergenic—people who have allergies are most likely not allergic to sheep.

Sheep for Milk

The dairy sheep industry is highly developed in and around the Mediterranean Sea. In the United States, sheep dairying is a small part of a small industry. Most US sheep dairies are located in the Upper Midwest and New England states. At the same time, there are sheep dairies in most states, and there is potential for expansion of the industry, as the majority of cheese made from sheep milk is imported.

In a dairy sheep operation, the sheep are usually triple-purpose, but the emphasis is on milk production. Management decisions are usually made to maximize milk production. Additionally, lamb sales can contribute a significant amount of income to the dairy sheep enterprise, especially if lamb prices are high or if breeding stock can be sold for premium prices. Wool receipts are usually negligible unless fleeces are direct marketed or value is added to the clip. Sheep dairying is sometimes combined with agrotourism.

Sheep milk is superior to milk from goats and cows for cheese-making. It is much richer, containing a higher percentage of fat, solids, and protein. For this reason, sheep milk gives a much higher cheese yield than the milk from cows or goats. Yogurt and ice cream are also commonly made from sheep milk. Sheep milk is rarely consumed as fluid milk.

Chapter 3: Facilities and Equipment

Housing for Sheep

Housing for sheep was covered on a small scale in the previous chapter. In Chapter 3 we are going to cover housing for small and larger flocks of sheep. There are three ways to think about shelters in terms of sheep: the first is to believe that less is more. If you have a hardy breed of sheep and you live in a climate that is not extreme during the winter or summer months, you probably do not need elaborate shelters.

The idea for building simple shelters stems from the thought that the fleece is insulating material for the sheep. However, you also have to consider when your ewes are going to lamb; you do not want your ewes out in the middle of freezing weather having lambs with little or no shelter. You at least need a place for your ewes to lamb and a warm place for lambs when they are young.

Another option for shelter is to build a series of smaller sheds so that the sheep can congregate together, but still have enough room to be comfortable. You could order a kit or construct one or two 12' by 20' sheep barns for a smaller flock of sheep if you live in a climate where a minimal amount of shelter is needed. Some people build pole barns for their sheep and herd the entire flock into the barn during the night or when there is extreme weather.

Image of a Port-A-Hut Shelter from
"A Beginners Guide to Raising Sheep"
Sheep 101[xvi]

Article with Instructions to Build
"A DIY Mobile Sheep Shelter"
Countryside Magazine Blog[xvii]

When creating living spaces for your sheep, it is important that you make sure all sheep have enough space. The recommended amount is a 20–25 square feet per sheep. However, some sheep need more and some need less. The recommended space outdoors is 1 acre for every three to six sheep. This land should be good grazing land.

You can have a full size enclosed pole barn for the flock to have access in case of inclement weather. The floor should have a dirt surface and the barn should be clean. This helps the animals stay cleaner and healthier. The dirt floor helps the sheep keep their joints from getting sprained.

Lesgets.com via Pinterest[xviii]

Feeding troughs

Having feeding troughs for supplemental food or for hay in the winter and/or summer is almost essential if you live in a place where weather experiences extremes. A feeding trough prevents waste and makes sure sheep are well-fed. You do not have to get fancy with your trough. The trough can have access to one or both sides. You should make sure you have 30 cm per lamb and 50 cm per grown sheep to have room to feed. The suggestion is less for double-sided but stick with these measurements and you will have happier sheep.

These self-feeders can save time and labor when sheep are unable to graze. Feeding troughs have to be raised off of the ground so the sheep do not get underneath the hay and overeat. Many shepherds make their own feeding troughs. Pinterest is full of ideas. The photo below is one example.

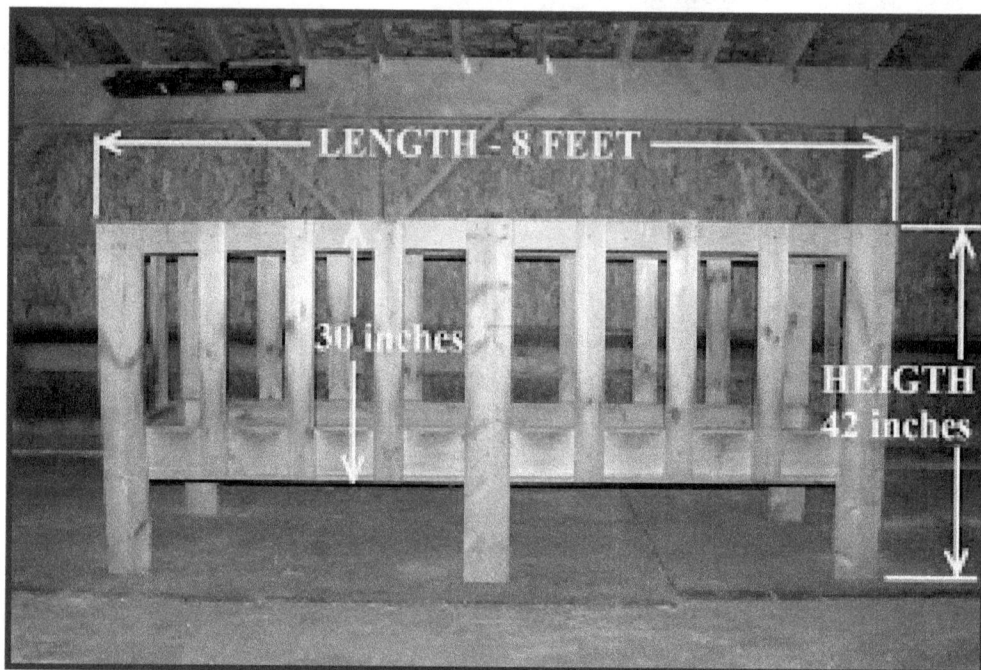

Ugly Dogs Farm
Blogpost that describes how to make this feeder[xix]

Fencing

Fencing has a two-fold purpose. You are trying to keep your sheep on your property and you also want to keep your sheep safe from predators. Fencing to do both may be expensive at first, but to have a safe and healthy flock, your fence will pay dividends down the road.

The Gallagher Smart Fence is a new product. This is a new and easy way to use temporary fencing. This is a portable fencing mechanism to segment your sheep paddocks for rotational grazing and to create temporary paddocks. You will combine this with a small portable electric fence charger and this will give you a fairly inexpensive sheep fence solution.

Welded Wire Fencing for Sheep

ProFence, LLC[xx]

This fencing can be purchased in rolls to fence one acre at approximately three rolls for just under $500. The wood posts will cost close to $350, and the steel t-posts will run close to $300. You can fence an acre for just over $1000. This does not include the steel gate or some of the bolts, etc. But as the picture shows, this fence is sturdy and will hold your sheep and discourage most predators.

Make sure the height of the fence is tall enough so that dogs and coyotes cannot jump it. Check the fence often during different seasons for gaps or weathering that may have caused damage. Other than that, this fence is effective and secure for a happy, healthy flock.

High Tensile Electric Fencing

High tensile electric fencing is an effective perimeter fence for your flock. This fence is able to carry a strong enough charge for the sheep and also to deter predators. This fence is fairly easy to maintain and makes it easy to make sections to rotate pastures. This fence may take longer to install but is easy to maintain and is cost effective. The cost per acre to set up this fence is just under $2000.

Gallagher Electric Fencing[xxi]

Electric Poly Rope Sheep Fencing

This fencing is effective for rotating grazing and is less expensive than other electric fences. However, the fence should be checked often because it is not as durable as fences such as the high tensile electric fence. This fence will cost about $1000 per acre.

W-Bar-Y-Fence[xxii]

Traditional Sheep Fence

Traditional sheep fences are usually for larger sheep farms. These fences include the barns or sheds and grazing land. They are made of wooden fence panels with lightweight tubular metal gates and panels. You can create smaller sections for making holding pens and small paddocks for flexibility.

Stop Gap Fencing[xxiii]

Plastic Mesh Fencing

For smaller flocks and also for rotating flocks, some shepherds use plastic mesh electrified fencing. This is highly visible and is effective to keep sheep in and predators out. When purchased, it comes with the stakes that help you install the fence fairly easily. The rolls come in 50 foot lengths. The price per acre is just under $1000. A drawback to this fencing is that it can get tangled during storage.

Chapter 4: Grazing and Nutrition

Grazing

If sheep have been cooped up during the winter months, then the grazing season is spring, summer, and fall. The benefits of grazing are:

- exercise and fresh grass,
- saving of expenses, and
- manure enriches the soil and helps build organic matter

Shepherds cannot wait to get their sheep out to graze, but there are proven steps to get the sheep out to the fields in the best way.

1. Before the sheep are sent out to graze it is time to shear them. After sheep are sheared, you will be able to evaluate how the sheep faired during the winter months. This is a good time to weigh and measure the herd.
2. Trim hooves and inspect feet to prevent accidents and sprains.
3. All sheep should be vaccinated and dewormed before being turned out to pasture. Sheep can become sick from overeating and these vaccines help with a number of situations.
4. Check all the fences and make sure they are going to hold.
5. Make sure the sheep have been fed with the hay you have in stock before you turn them out. If they are hungry, they might eat too much too fast of fresh grass and that wouldn't be good for their stomachs.
6. Make sure there are open sheds where the sheep can go in case of storms or late spring snows. Also, they should have protection during hotter days.
7. Provide an adequate supply of clean water for your flock.
8. The chosen pasture should have good length of grass levels. If you have tall fescue or brome grass, provide sheep with magnesium supplements to help prevent grass tetany. If your pasture has alfalfa, limit grazing in that area to prevent bloat.
9. Have a good predator system in place. Llamas, donkeys, and dogs are all great for keeping predators at bay. More details about predator systems will be in another chapter.

Different Grasses

Clover

Clover is a preferred grazing grass for sheep. Clover has the advantage of grabbing nitrogen from the air. A percentage of clover higher than 30% in the pasture takes care of the needed nitrogen for that pasture. A pasture with 50% is even better.

Orchard Grass

Orchard grass is one of the highest yielding grass species. It is drought resistant. Sheep will eat it and will not graze this grass too short. This grass is usually left over for the late summer and fall seasons. Sometimes there is grass left during winter.

Fescue Grass

Often the sheep are displeased with some types of fescue. Meadow fescue seems to be the most popular with sheep.

Bluegrass

Bluegrass is a favorite for most sheep. This is a persistent grass. It grows high and mixes well with other grasses.

Baraula Grass

This grass has a high yield without too much care. Sheep love Baraula grass. This grass also stockpiles well for winter months.

Nutrition for Sheep

It is of the utmost importance for sheep to have the correct nutrition in order for them to have the best production records. It is also important to have a happy flock. You will have happy sheep if you have adequate nutrition and water.. Happy sheep equal a happy farm.

Energy for sheep, or calories they intake, for the most part comes from their pastures. If your pastures have adequate grazing grasses and roughage, your sheep will have adequate energy. However, if your pastures are lacking and the sheep are grazing down too far, the energy input may be too low for your sheep. A deficiency will result in reproductive failure, decreased production, higher mortality, and your sheep may be more susceptible to parasites and disease. It is important to make sure your portions of grazing land are enough per sheep. You should measure your total digestible nutrients (TDN) to make sure every sheep is getting what they need to be healthy, productive animals.

Protein is particularly important for lambs when forming bones and muscles. When adding protein to the diet of sheep the quality is more important than the quantity. The most common protein supplement for sheep is soybean meal. This may be the most expensive addition to a lamb's diet, but it is very important.

Minerals (Virginia Cooperative Extension)

There are many minerals that are required in the diet of sheep. **Macrominerals** are required in larger amounts, with that requirement expressed as a percent of the diet or as grams per head per day.

- **Adequate**   Potassium
- **Deficient**   Sodium (when combined with Chlorine, makes salt)
- **Marginal**   Calcium, Magnesium, Phosphorous, Sulfur

Calcium is often found in adequate amounts in forages, and legumes have higher levels than do grasses. Grains and grain crop silages have very low levels of calcium. Phosphorous is just the opposite. It is high in grains and low in forages, often because soils are low in phosphorous fertility levels. Because phosphorous is important to reproduction and growth, it is often included in minerals for the ewe flock year around. Magnesium is often low in lush forage growing in early spring or when spring-like conditions occur. A deficiency of magnesium causes grass tetany, a problem in cows that rarely occurs with ewes.

Microminerals

Minerals needed in very small quantities are called microminerals, or trace minerals. The requirement by animals for these minerals is expressed in milligrams per head per day or in parts per million. Just as with the macrominerals, some are adequate, others are deficient, and some are marginal.

- **Adequate**   Manganese, Iron
- **Deficient**   Selenium
- **Marginal**   Zinc, Copper

Iron is often added to minerals (iron oxide or ferric oxide on the tag), even though the required amount is included in the forage that is consumed in the basal diet. The reason it is added is to give minerals the typical reddish-brown color. However, iron can interfere with the uptake of other minerals that are not in large amounts, such as zinc. Thus, it is recommended that iron not be included/added to complete minerals for ruminants.

Zinc, Copper, and **Selenium** are all important in many physiological functions, including the immune response and disease-fighting ability. Our soils are often deficient in Selenium, making forage grown on those soils also deficient. Consequently, it is strongly recommended to include Selenium in mineral mixtures for sheep of all ages.

<u>Salt</u> is also essential for sheep to have daily. Inadequate salt intake can decrease water and food intake, milk production, and the growth of lambs. You can tell if your sheep are salt deficient if they are chewing on wood or licking dirt. Some sheep will even begin to eat rocks. They might be more likely to consume poisonous plants. Having salt blocks available is a good way to make sure sheep are getting salt every day if you think they may have a shortage. Sheep will go for the blocks.

<u>Water</u>

All living things need water to survive. The question is whether sheep will need supplemental water during winter, for example, as well as other seasons during the year. You will know based on the following factors if your sheep are lacking water:

- Are your sheep productive? Do the ewes nurse properly and are lambs growing?
- Check environment temperature and humidity.
- What is the water content of their feed and grass they consume?
- How much consumable water is available through snow?

The sum total of these conditions will tell you if your sheep need supplemental water. If this is not your first rodeo, then you know when your sheep need water. But believe it or not, there are some farmers who overwater the animals and this can cause a bloating situation. Many sheep will keep drinking if water is available, whether they need it or not. If you are unsure of when to give sheep supplemental water then follow this guide. Each section is described more in depth.

Productive State of the Animal

Mature lambs and ewes at **maintenance level** have a lower water need in comparison to when they are at other productive levels. Maintenance level is when ewes are not carrying lambs or nursing. They do not need as much nutrition as they do during the more productive times. If an ewe is nursing, that animal needs a high level of water. Not all sheep need the same amount of water every day. An ewe that is in the first 30 days of lactation needs more water than the end of the lactation period. The more lambs the ewe is nursing, the more water she will need.

Environmental Conditions

If the temperatures are mild, between 35- and 72-degrees Fahrenheit, then the sheep are going to need 50% less water than if the temperatures are much above or below those numbers. Water needs of sheep are usually less during winter unless the temperatures are extreme. In many spring, fall, and winter grazing situations, their water needs can be entirely met with water contained in grazing forage and/or from soft snow.

In grazing situations, relative humidity also plays a factor, because the temperature at which dew accumulates on grass is directly related to the humidity. During fall and spring grazing, when the temperature begins to drop and the air is humid, there will be abundant dew on the grass which can provide adequate water to grazing animals. Beginning in September, sheep will most likely ignore water troughs if they are getting enough water from foraging in the pasture. Summer is the critical time to put the water troughs to use unless you live in an area where your flocks will get an abundance of rain.

Water Content of Feed

If sheep are housed indoors in extreme weather situations and are fed dry food, they are going to need supplemental water. Regardless of temperature, if sheep are receiving dry food, water supplements will be needed. This is different, as you read, when sheep are grazing. Regardless of dry feed or grazing, ewes that are pregnant or lactating should always be given supplemental water.

Nutrition and Weaning Lambs (Penn State)

Lambs should have all their vaccinations prior to weaning. They should receive their first vaccination shot within the first month after they are born. A booster vaccination should be given three to four weeks later. Deworming, castration, and tagging should also be completed well ahead of weaning as all these practices cause additional stress to the lambs at weaning. A lamb needs the comfort of its mother after suffering through vaccines and other procedures, so it is important to do these things while lambs are still connected to their mothers.

If unable to perform these tasks prior to weaning, producers should wait until several weeks after weaning when lambs are well adjusted to living without their mothers. Watch lambs closely for any signs of health problems for several weeks after weaning. This includes problems such as pneumonia, scours, and coccidiosis.

The most common form of weaning is to let the weaning process happen naturally in the pasture from ewe to lamb. Shepherds who are producing sheep for wool are most likely to prefer a natural weaning method where lambs stay on pasture with their mothers until they are four to six months old. Stress is greatly lowered by this method for both ewes and lambs and people who have to listen to them. At any rate, lambs should be 45 pounds before they are weaned. If lambs are in a grazing flock, they should be full on grazing and eating independently by the time they are weaned naturally.

Chapter 5: Shepherding Skills

Training Yourself to Be a Shepherd

Training yourself to become a shepherd isn't difficult. You should start with a smaller flock and then if you are successful raising a few sheep from lambs to adulthood and through a reproduction cycle, you might be ready for more if you like the process so far. Once you have completely read this book, you will have all the information you need to raise a flock of sheep and you should always consult with your local veterinarian if in doubt about a sheep or a lamb. Other shepherds are also good resources as well.

Be patient with yourself and with the sheep. Start with no less than five sheep because sheep like being part of a herd. For the most part, sheep also like people. Go outside and spend time with your flock.

1. **Shelter** – You already know from what you have read so far in this book that you need shelter for your sheep. Refer back to Chapter 3 in order to decide what kind of shelter you need for the size of flock you have. It is recommended you start off with a smaller flock if you have never raised sheep before.
2. **Appropriate fencing** – Put appropriate fencing around the grazing area and including the shelter. Refer back to Chapter 3 of this book. It is complete and will also provide cost per acre.
3. **Feeding** – Allow grazing time and, if there is not enough grazing, supplement hay. A handful of grain is enough to supplement. Too much grain can cause problems for the sheep. Don't fall for the Labrador dog trick. Sheep always act hungry. Look at them. If they are filled out, they are fine. Just like a Lab.
4. **Make sure their pens are clean** – Dirty pens are breeding grounds for insects, worms, disease, and parasites. The worst is flystrike. This will be discussed in a later chapter. Rake out any old hay that is wet or has feces. Replace with dry, fresh hay.
5. **Water** – Make sure during summer months supplemental water is fresh and clean.
6. **Grazing** – Sheep can be allowed to graze all day as long as you have adequate fencing and shelter.

How to Handle Sheep

The more used to you the sheep are, the easier they are to handle. Most sheep dislike being handled very much. They become almost offended that someone wants to handle them at all, no matter who it is. Do not try to handle sheep unless you are calm. They are just like teenagers and can read the room. If they see you sweat, it is over. You have to be in a place of Zen to be able to get the upper hand with your sheep.

Sheep are defensive. It is their most valuable gut instinct that keeps them alive. If you panic, they will panic. Shearing during a state of panic is never good. Before this situation gets revved up into that state of panic, you have to take charge. The first step is to create a holding area. This is going to be a small space for a small amount of sheep. An area measuring 5 x 5 or 6 x 6 is perfect for a holding area. Sheep will have no room in a very small area to gain speed on you.

If you have movable livestock panels, this will work best when gaining ground on one sheep. One or two panels should be carried as you move toward one sheep and walk that sheep into a corner. Once the sheep is in the corner, you can use the two panels to box the sheep into its own very small pen.

It is not best practice to grab a sheep by the horns. Try to cup one hand beneath the sheep's jaw. Use the other hand to press against the sheep's back or shoulder. When you cup your hand under their jaw, they will start to back away. Do not let them. Use your knee to push them up against the wall to examine them or to inject with necessary medicine, etc., or to inspect the sheep for other matters. This is also when the sheep is put into the shearing area for quick shearing.

Shearing Sheep

For small flocks of sheep, many farmers shear themselves. Some will still hire this task out to professional sheep shearers. A professional will shear sheep at a rate of two minutes each and will get the wool off of the sheep in one piece as a rule. An inexperienced person cannot do that. Regardless of how the wool is used, sheep must be sheared every year in the spring because wool does not stop growing. Shearing sheep is a very labor-intensive task that requires practice and skill. For the most part, electric shears are used to shear the sheep.

When a sheep is being sheared it is important not to cut the hide because if the hide is damaged, it can become infected and wool may not grow in that place again. The biggest problem is that cutting the hide causes the animal to feel pain that doesn't have to happen during shearing.

If you decide to shear yourself, wear lightweight shoes or moccasins if you have just a few sheep. Lightweight shoes with rubber sheepskin bottoms will keep you from slipping. The lighter the shoe, the less weight you have on your feet as well. Shearing is a very tiring activity. A good place to lose the weight is by wearing lighter shoes. Also, you won't sweat as much. Most of the time you will be maneuvering sheep on the ground. If you have a large flock, you should hire professionals to help. This will expedite the shearing process.

When you are shearing the sheep yourself, you will start by shearing an ewe's chest, belly, and crotch while the sheep is sitting up on its bottom. Next allow her right front leg to slip back between your legs and use your knees and feet to turn her onto her right hip. This puts her into position for you to shear the whole of her left hind leg right up to (and preferably a little over) her backbone, and her tail. Having done this, you then step forward while lifting the sheep up into a higher sitting position in order to shear the left hand side of her neck. Then you will shear the left shoulder followed by long stretches of the left side. Then you will start on the right side in the same way. You will finish by shearing the legs.

Flystrike

If sheep are not sheared properly in the spring, the infection *flystrike* can infect them. Flystrike is another word for a severe infection of maggots. This is one of the nastiest infections sheep can get. Flystrike is caused by the greenbottle fly. This fly will lay its eggs in damp and soiled areas of wool on the back ends of the sheep. Sometimes they might lay eggs in bad hooves, and then the maggots will spread to the side of the sheep. When the maggots hatch, they will eat the animal alive if left untreated. For this reason, sheep must be inspected often. Shearing sheep in the spring is extremely important.

You might become aware of flystrike if the sheep becomes lethargic, but you should notice this horrible infection before it gets that far along. An animal may be chewing on itself or licking incessantly. If the sheep has light colored wool, the patch may be easy to spot, but if the sheep's wool is dark in color, spotting the sore area may be almost impossible to see. If flystrike is caught early, it is easy to treat, but if it is caught too late, the sheep may die.

Crookwell Veterinary Hospital
Flystrike, Prevention and Treatment[xxiv]

Halter Training Sheep

Just like some dogs, sheep will be much easier to handle when they are halter trained. Halter training a sheep is worth the effort if you work with sheep on a smaller farm. With a halter, you can walk your sheep to a fence or a holding area and attach a short leash to the side of the area and work on the sheep while they are in a standing position.

In the beginning of training, offer oat treats when the sheep is standing or walking nicely using the halter. The idea is to get sheep used to the halter when they are very young so they know this is fun. They are going for a little walk, getting treats and being spoiled, essentially before it is actually necessary to do any procedures or shearing by way of leading with the halter. Do not ever leave the halter and leash on any sheep when you are not right next to that sheep. They can get tangled up in ways you cannot imagine and choke or break a leg.

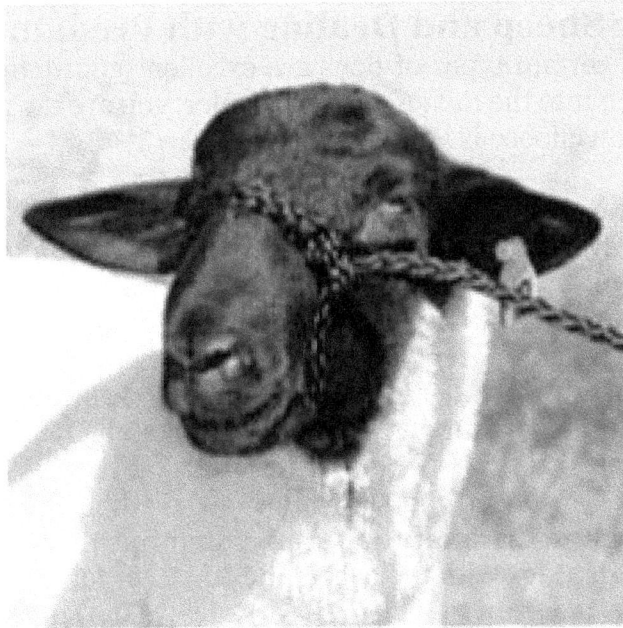

ValleyVet
Poly Sheep Halter[xxv]

Modern Things Sheep Ranchers Advise

1. Most sheep ranchers would rather be called ranchers or flock owners instead of shepherds.
2. Sheep are smarter than you think. If you underestimate their intelligence, they may outsmart *you*. Sheep have survived for thousands of years. They must be doing something right.
3. Keep your eyes on the flock, but care for each individual sheep. If you know each sheep and take care of each sheep, your whole flock will be easier to care for. Take the time to get to know each sheep and for them to know you. This will pay dividends down the road.
4. Make sure you are there when an ewe gives birth. Sometimes a shepherd must assist in order to save the ewe's life or the life of the lamb or both. Know your sheep well enough to know when an ewe is going to deliver. The miracle of watching and helping in the birth process is something you do not want to miss. Seeing the mother care for her newborn lamb is nothing short of amazing.
5. Baby lambs are cute and soft. Hold them when you can and get them used to you early on.

Guardians for Sheep and Dealing with Predators

Llamas and certain types of dogs are excellent guardians for sheep. Even donkeys have gotten into the mix. As a sheep owner, you are the primary caretaker, but the guardian you choose is your best friend.

Llamas as Guardians

Often you will see Llamas and dogs working together in the fields to protect the flock. Llamas have keen eyesight and are able to spot anything out of the ordinary quickly, such as a strange dog or coyote skulking around. They react and will run toward the predator at high speed with their head down and stretched out in order to head butt the intruder. Llamas will send the unwanted animal rolling. Usually, a coyote cannot get out of the field quick enough.

Llamas make an unsettling screaming noise when they get upset, which sometimes is enough to drive off the stranger and put the sheep on high alert. Llamas stamp and also kick with their feet. They do their job very well. Guard dogs and llamas also play tag team in the chase with stray dogs and coyotes.

When sheep are being attacked, they will circle up and bunch together, making them easier to protect. The llamas and guard dogs can then stay on the outside of the circle and protect the flock much easier when they are tight together.

Most coyotes and wild dogs have no idea about how fierce the llama is, or for that matter, what the llama is, and that fact alone usually sends them running. However, llamas have been known to stomp a coyote to its complete demise.

Farming with Carnivores Network
"Guardian Llamas"[xxvi]

As a rule, llamas like people unless they have a bad memory of someone in particular. Llamas are intelligent and will get to know their family, flock, and other guardians very quickly.

Guardian Dogs

When you picture a guardian dog you almost always picture the black and white Border Collie. If you know very much about Border Collies, you know their disposition is very sweet and gentle. However, they do nip without hurting the back legs of a sheep to get them to do what they want them to do.

Would a Border Collie win a fight against a wolf or a coyote? Probably not, but because the sheep are a part of the Collie's pack, the Collie will arch her back and bark like she will go into full attack mode. This is usually enough to scare off any predator.

The Spruce Pets
Blogpost "7 Kinds of Working Dogs"[xxvii]

Another common sheep dog is the Great Pyrenees. This dog breed is large, white, and has big floppy ears. From a distance you cannot tell these dogs from the sheep. At first glance they don't appear to be fierce, but again, do not mess with their "pack," as they are very territorial as well. Usually, their bark and stance will scare away another dog or other type of intruder.

Wide Open Pets
Sheep and Pyrenean Mountain Dog[xxviii]

Having your guard animal in place will be a big step to help keep predators away from your sheep. If you have a large flock and several acres, checking your fencing often is another way to help ensure you do not have any unwanted visitors. Provided your fencing is in good condition and you have all of your guardians in place, you should be able to feel secure about your sheep being in the fields overnight in nice weather. Sheep should always be able to get to shelter if need be if sudden storms should blow in.

Chapter 6: The Business of Sheep for Meat

Lambing Management

Lambing is the process of raising sheep for meat production. The key is to raise the lambs to be a certain age in order to have the highest quality of meat. In order to raise sheep for meat production, management is key. The larger the flock, the more organized the system needs to be. Lambing in and of itself is when ewes have lambs.

Profitable sheep production requires the application of certain well-timed management practices to ensure the overall well-being of the flock. Advances in breeding, lambing, feeding, and health management have given producers the tools to increase both the number and weight of lambs marketed annually.

No single system of production is right for everyone. The ideal lambing season for a flock will depend on the available facilities, labor and management resources, genetics, pasture and feed resources, and the operation's marketing program. However, each system of production must emphasize those practices which enhance the overall well-being of the flock. Areas of critical importance that must be considered for every flock include internal parasite control, foot rot control, and predator control.

The following schedule is divided into three parts. This schedule is provided by *Virginia State University* and is available for public use. Full credit will be given at the end of this chapter and in our sources section at the end of this book. Section 1 of the schedule includes jobs that should be performed during particular seasons of the year regardless of lambing date. Section 2 lists management practices that should be performed according to breeding and lambing dates. Section 3 gives some general recommendations that should be considered as they relate to overall flock health, facilities, breeding, and feeding management.

Section I
Winter

1. Ewes up through 15 weeks of gestation should receive 4 lbs. of a good quality grass/legume hay daily.
2. Ewes in the last 4 weeks of gestation should receive 4 lbs. of a good quality grass/legume hay plus 1 lb. of corn daily.
3. The highest quality hays should not be fed during gestation. Utilize average-to good-quality hays during the early gestation period, when ewe nutrient requirements are low compared to late gestation and lactation. If high-quality hays, such as alfalfa, are fed during gestation it is important to limit intake as overfeeding is costly.
4. If hay is limited, 1 lb. of corn may be substituted for 2 lbs. of hay. To prevent wool picking and digestive disturbances, ewes should receive a minimum of 1.5 lbs. of hay per day.
5. Provide fresh drinking water, free of ice, every day. Feed intake is severely depressed for sheep deprived of water for more than 24 hours. Lack of water predisposes ewes in late gestation to pregnancy disease.
6. To minimize feed wastage and to avoid the spread of disease, hay and grain should not be fed on the ground.
7. Provide a complete mineral mix free-choice. The mineral should be specifically formulated for sheep and fortified with selenium.
8. Lactating ewes with singles should receive 5 lbs. of a good quality grass/legume hay plus 1 lb. of a 15% crude protein grain mix daily.
9. Lactating ewes with twins should receive 5 lbs. of a good quality grass/legume hay plus 2 lbs. of a 15% crude protein grain mix daily.
10. Avoid damp, dark, or drafty barns, and wet muddy areas in or around buildings. Young lambs are able to withstand cold temperatures quite well, but drafts and dampness can lead to losses from baby lamb pneumonia.
11. Market spring-born lambs that have been grazed through the summer and fall. Grain supplementation the last 30 to 60 days prior to marketing will provide efficient weight gain and enhance market readiness. Spring-born lambs should be sold no later than April.

Spring

1. Treat ewes and all lambs over 6 weeks old for internal parasites the day before turning onto spring pasture. Thereafter, treat lambs and ewes once every 3 to 4 weeks throughout the grazing season.

2. Shear during March, April, or May if ewes were not shorn pre-lambing. Shear white face ewes first and package and market their wool separately from blackface and blackface cross ewes.
3. Store wool in a clean, dry place. Do not store wool on the ground or on concrete.
4. Sell cull ewes in late February, March, or early April.
5. Trim and check feet and put the flock through a foot bath prior to placing ewes on pasture.
6. Move ewes nursing lambs to fresh pasture every 2 to 3 weeks. An ideal sheep pasture contains a mixture of grass and clover with an average height of 4" to 6".
7. Restrict the grazing flock to two-thirds of the available pasture. Harvest hay from the remainder. Allow time for regrowth and graze for the rest of the year.
8. Identify and retain ewe lambs from a winter lambing to be used as replacements. Breed so that they will lamb first as yearlings.
9. Manage winter-born lambs (December through March) so that they are marketed in the spring prior to the season price decline that occurs heading into the summer months.

Summer

1. Provide a complete mineral mix, specifically formulated for sheep and fortified with selenium, to the flock during the grazing season.
2. If adequate pasture is available, allow spring-born lambs to graze on the ewes into the fall. Allow spring-born lambs that will be marketed in late fall and winter to attain cheap weight gains from forage. Spring-born replacement ewe lambs may be weaned, sheared, and placed on higher-quality pastures to optimize development prior to the breeding season.
3. Treat ewes and lambs for internal parasites at least once every 3 to 4 weeks. Failure to do so will result in poor lamb performance and unnecessary loss.
4. Starting in August, stockpile fescue pastures for late fall and winter grazing.
5. Assess ram inventory to decide if additional or new rams need to be acquired and investigate potential sources of new rams. Acquire replacement ewe lambs if open females are to be brought into the flock prior to the breeding season. Isolate sheep from outside sources a minimum of 4 weeks before placing with the existing flock.

Fall

1. Test hay and silage samples to determine their nutritive value. Work with an Extension agent to determine the supplements that will be required to formulate balanced diets for winter feeding.
2. Trim and check feet.
3. Graze spring-born lambs on available fall pasture and aftermath hay fields.
4. Avoid marketing lightweight lambs in September and October when market prices are typically at their lowest levels of the year. Market spring-born lambs in late fall through winter.
5. Supplement grain on pasture to promote increased lamb weight gains.
6. Identify and retain ewe lambs from spring lambing to be used as replacements. Breed so that they will lamb first as yearlings.
7. After November 1, place ewes on stockpiled fescue pasture. With adequate fall growth, one acre of stockpiled fescue should supply enough feed for five non-lactating ewes through February 1. Use temporary electric fence to limit the sheep's access to a portion of the stockpiled pasture until fully utilized.

Section II
Six Weeks Before Breeding

1. Wean all lambs that remain with the flock.
2. Catch and check all ewes that are limping. Trim their feet and treat if necessary.
3. Acquire replacement ewes, if replacement ewes will be brought in prior to the breeding season. Keep them isolated from the regular flock for at least 4 weeks to reduce the risk of bringing in diseases such as soremouth and foot rot.
4. Determine if a sufficient number of rams is available for breeding. A ram to ewe ratio of 1:25 for ram lambs and 1:35 for mature rams is generally recommended.
5. Have a veterinarian perform a breeding soundness examination on all rams. The breeding soundness exam should include semen evaluation.
6. Condition score rams both visually and by handling them down their top. Rams in thin condition should receive 2.5 lbs. of grain per day in addition to their normal diet, while rams in moderate condition should receive 1 lb. of grain daily.
7. Shear the rams. Rams becoming overheated or running a fever as a result of sickness may be sub-fertile for much of the breeding period.

8. For spring and summer breeding, keep rams out of sight and sound of the ewe flock until the first day of breeding. Avoid fence line contact. Breeding performance will be improved as a result of the "ram effect."
9. Vaccinate ewes for abortion diseases as per label indications.

Two Weeks Before Breeding
1. Treat ewes and rams for internal parasites.
2. Place ewes on high-quality pasture. Avoid pastures with 50% or greater stands of clover or other legumes. Legumes have been shown to contain estrogenic compounds that lower conception rates.
3. Flush ewes by feeding 1 lb. of whole corn or barley per head daily, starting 2 weeks before the breeding season and continuing 2 weeks into the breeding season. The practice of flushing improves lambing percentages by 10% to 15%.
4. Keep unfamiliar rams together in a small pen for 3 to 5 days so they will become accustomed to one another. This prevents death or injury that could occur from fighting.

At Breeding
1. Where practical, provide the breeding flock access to a cool barn, shed, or woods during the hot part of the day.
2. Use a marking harness on all rams. This helps to determine the percentage of ewes that are cycling and helps to evaluate the breeding performance of the rams. Change colors on the harness every 17 days. A large number of ewes re-marking may indicate a ram or ewe fertility problem. By recording breeding dates on the ewes after they're marked, they can be sorted and managed more appropriately for lambing.
3. When using more than one ram in a group of ewes, try to use rams of similar size and age. Ram lambs should not be used in multiple-sire breeding groups with mature rams. Larger, older rams tend to dominate smaller rams and breed more than their share. This may result in lower conception rates and lower lambing percentages.
4. Divide the flock into single-ram units. A ram to ewe ratio of 1:25 for ram lambs and 1:35 or more for mature rams is generally recommended. Change or rotate rams every two weeks and observe the flock closely to be certain ewes are settling. Exercise a controlled breeding season and remove rams after 60 days.
5. Rams can lose up to 12% of their bodyweight during a 45-day breeding period. Be prepared to supplement their diet with grain whenever possible.

6. Provide a mineral supplement specifically formulated for sheep on a free-choice basis throughout the breeding period.

Breeding to 6 Weeks Before Lambing
1. Mature ewes in average to good body condition should be fed to maintain or slightly increase their bodyweight during the first 3.5 months of gestation. This is the time to take advantage of poor quality pasture or crop residue. If this period occurs during the winter, hay or silage will do the job, with no supplemental grain required.
2. Thin ewes should be fed separately and supplemented with 1 to 1.5 lbs. of grain per day to gain 10 to 15 lbs. by 6 weeks before lambing.
3. Pregnant ewe lambs should be fed separately from mature ewes. They should gain approximately 25 lbs. from breeding to 6 weeks before lambing. Attempts to cause large weight gains in ewe lambs during late gestation may lead to lambing problems.
4. If pregnant ewes are to be brought into the flock, keep these ewes separate from the main flock through lambing when feasible. This will diminish the risk of introducing abortion and other diseases into the main flock.

6 Weeks Before Lambing
1. Start feeding 0.5 lb. of grain per head daily as a preventative for pregnancy disease. Grain may be in the form of whole shelled corn or barley. Even if ewes are on good quality pasture, they still require the extra grain. During the winter or when on poor quality pasture, feed approximately 4 lbs. of hay in addition to grain.
2. Supplementation of tetracycline pre-lambing has been shown to reduce the incidence of abortions. Consult with your veterinarian on a flock health management protocol.
3. Make sure there is plenty of feed trough space so that ewes do not crowd each other at feeding time.
4. Check and avoid ditches, sills, narrow gates, or any other objects that would cause ewes to jump, crowd, squeeze, or climb before lambing.

4 Weeks Before Lambing
1. Shear the wool from around the head, udder, and dock of pregnant ewes. If covered facilities are available, shear the ewes completely. Sheared ewes are more apt to lamb inside. The inside of the barn stays drier because less moisture is carried in by the ewes, more ewes can be kept inside, and it creates a cleaner environment for the lambs and the shepherd. Sheared

ewes must have access to a barn during cold, freezing rains, and they must receive additional feed during periods of extremely cold temperatures.

2. Vaccinate ewes for overeating disease and tetanus. These vaccines provide passive immunity to baby lambs through the ewes' colostrum until they can be vaccinated at 4 to 6 weeks of age.

3. Check and separate all ewes that are developing udders or showing signs of lambing. Check and remove heavy ewes once a week during the lambing season. Increase the grain on all ewes showing signs of lambing to 1 lb. daily and feed all the good quality grass/legume hay they will clean up.

4. Observe ewes closely. Ewes that are sluggish or hang back at feeding may be showing early signs of pregnancy disease. If so, these ewes should be drenched with 2 ounces of propylene glycol 3 to 4 times daily.

5. Shelter heavy ewes from bad weather.

6. Get lambing pens and lambing equipment ready. There should be one lambing pen for every 10 ewes expected to lamb.

7. Stock lambing supplies such as iodine, antibiotics, frozen colostrum, stomach tube, injectable selenium and Vitamin E, OB lube, lamb puller, ear tags, etc.

At Lambing Time

1. Check ewes on a frequent basis (every 3 to 4 hours), as feasible. Do not check ewes in the middle of the night. Activity at that time may stimulate ewes to lamb two to three hours before they normally would.

2. Lambing cubicles placed around the walls in the lambing area of the barn measuring 4' X 6' have been used successfully as a place for ewes to lamb away from the other ewes in the barn. The cubicles have a 2' wide opening with a 10" board as a threshold to keep lambs inside.

3. After lambs are born, move the ewe and her lambs to a lambing pen with a minimum dimension of 5' X 5'. Check the ewe's udder to see that she has milk, strip each teat to remove the waxy plug that may be present at the end of the teat, and make sure lambs nurse within 30 minutes.

4. Colostrum is critical for baby lamb survival. For ewes without milk or for lambs that fail to nurse, lambs must be given colostrum via a stomach tube. If sheep colostrum is not available, cow or goat colostrum should be used. Colostrum can be frozen in ice cube trays or stored in "zip-lock" storage bags. Colostrum should be thawed using indirect heat. Thawing by direct heat destroys the antibodies that are present. Lambs should receive 20 ml (cc) of colostrum per pound of body weight. It works best if feedings can be 4 hours apart.

5. Only use a heat lamp if lambs are weak and chilled. Avoid danger of fire by hanging heat lamps 3' above the bedding and in the corner of the lambing pen. Block off the corner so that the ewe cannot get under the lamp.
6. Check on the health of the ewe and her lambs at least three times daily. Lambs that are lying down should be made to get up. Those that fail to stretch after getting up may have a problem that requires further examination. The biggest cause of baby lamb mortality is starvation.
7. If selenium deficiency has been a problem, lambs should be given an injection of 0.25 mg selenium per 10 lb. of body weight immediately after birth. A good quality mineral provided to the ewe flock on a year-round basis has been shown to be the best way to prevent selenium deficiency.
8. A general rule of thumb is for the ewe and her lambs to stay in the lambing pen one day for each lamb. Weak or small lambs may require a longer stay.
9. Ewes should receive fresh water and high quality hay the day of lambing. Don't feed grain until the second day. One pound of grain plus 5 lbs. of good quality hay will take care of their needs until moving to a mixing pen.
10. If ewes were not treated for internal parasites within 3 weeks of lambing, they should be treated prior to removal from the lambing pen.
11. Keep records on all ewes, noting those that had problems. Individually identify lambs so they can be matched with the ewe. The ability to match ewes and lambs is important to monitor performance, and individual identification is critical for making selection and culling decisions.
12. Move ewes and their lambs from lambing pens to mixing pens. Make sure lambs are matched up well with their mothers before moving to larger groups. Ewes with twins should be receiving 2 lbs. of a 15% crude protein grain mix and 5 lbs. of good quality hay daily. Ewes with singles should be receiving 1 lb. of a 15% crude protein grain mix and 5 lbs. of good quality hay daily.
13. All lambs should be docked and castrated by the time they are 2 weeks old.
14. Lambs on a winter-lambing program should have access to a high-quality creep feed by the time they are 7 days old. Creep feeds should contain 18% to 20% crude protein and be low in fiber. Make sure the source of protein in commercially prepared lamb creep pellets is all natural protein and does not contain urea. Maintain at least a 2:1 calcium to phosphorous ratio in the feed by adding 1% feed grade limestone. Calcium to phosphorous ratios of less than 2:1 may lead to urinary calculi. When constructing a creep area, keep the following points in mind: 1) place the creep in a convenient location close to an area where the ewe flock congregates; 2) have openings on at least two sides of the creep and several openings per side; 3) keep the creep area clean and well bedded; 4) place a light over the creep to help

attract lambs—sunlight shining into the creep area works well; 5) keep feed fresh and provide clean water in the creep; and 6) construct the creep feeder so that lambs cannot stand and play in it. Allow 2" of trough space per lamb.

Section III
Breeding Stock

Ewes – Use only crossbred ewes for commercial sheep production. Crossbred ewes wean more pounds of lamb than the average of the purebred ewes that make up the cross. Crossbred lambs are more vigorous at birth and are heavier at weaning. Studies have shown that two-breed cross ewes mated to a ram of a third breed wean approximately 35% more pounds of lamb per ewe mated than the average of the purebred ewes producing purebred lambs.

Large differences exist between breeds for several economically important traits. For commercial flocks, it is unlikely that any one breed can meet production goals as effectively as a combination of breeds used in a planned mating system. Breeds need to be selected that contribute positively into a designed production system. Traits important for ewe breeds in crossbreeding programs include early puberty, moderate mature size, high fertility, optimum milking ability (appropriate for feed resources), longevity, management ease, and acceptable growth characteristics. Traits important in selecting a ram breed for use in crossbreeding programs include high growth rate with acceptable mature size, lamb survivability, and carcass merit. Retain multiple-birth ewe lambs for replacements that are above average in growth rate and born early in the lambing season. Breed replacement ewe lambs to lamb first as yearlings, and market open ewe lambs.

Rams – Buy sound, large, healthy, heavily muscled rams for market-lamb production. Sources for breeding rams include the Virginia Ram Performance Test, reputable consignment sales, or buying directly from a breeder. Regardless of the source, rams should be sold as guaranteed breeders. Purchase rams from sources that offer performance records and information that will be useful in making selection decisions.

Breeding Season – Available labor, barn space, weather, predators, lamb markets, and the amount and quality of feed and pasture should all be considered in determining the most appropriate lambing season. Fall and winter lambing are best suited for farms with good winter feed and suitable facilities, and for areas with high summer temperatures. Spring lambing is the preferred production system in more mountainous areas and has been shown to be consistently more profitable than other systems of production. Large operations find it best to breed ewes in groups and spread their lambings over a period of several months. Sheep are seasonal breeders. Most breeds and their crosses begin to cycle in late summer and are most fertile in the fall. On average, ewes exhibit heat every 17 days during the breeding season, stay in heat for 18 to 40 hours, and ovulate at the end of heat. The gestation period for sheep ranges from 140 to 159 days, with an average of 145 days.

Lamb Marketing – The ideal type, quality (grade), and weight for market lambs will be determined by the intended market and season of the year. Lambs may be effectively marketed at weights ranging from 50 to 120 pounds. Historically, lamb prices are seasonally highest in April and May, and lowest from September through November. Consequently, winter-born lambs should be marketed in the spring and early summer and spring-born lambs marketed in late fall and winter.

Virginia Cooperative Extension materials are available for public use, reprint, or citation without further permission, provided the use includes credit to the author and to Virginia Cooperative Extension, Virginia Tech, and Virginia State University. Scott P. Greiner, Extension animal scientist, Virginia Tech Publication originally written by Steven H. Umberger, Extension Animal Scientist.

Edwin J. Jones, Director, Virginia Cooperative Extension, Virginia Tech, Blacksburg; M. Ray McKinnie, Administrator, 1890 Extension Program, Virginia State University, Petersburg. May 2009.

Chapter 7: Vaccinations and Healthcare for Sheep

Suggested Vaccinations

1. Clostridium perfringens Type C & D (overeating disease) and tetanus – Vaccinate with an initial series of two injections administered 30 days apart. Thereafter, booster ewes annually at 4 weeks before lambing.
2. Vibriosis and EAE – Vaccinate with an initial series of two injections followed by an annual booster. Follow product label indications for the administration of the vaccinations relative to the breeding season.
3. Contagious ecthyma (soremouth) – This vaccine is not recommended if soremouth has not been diagnosed on the farm. For flocks that have soremouth, new flock additions that are brought in should be vaccinated, particularly if they have not previously been exposed to the disease.

Lambs

Clostridium perfringens Type C & D – Vaccinate with an initial series of two injections administered at 4 weeks of age and 1 week prior to weaning. Thereafter, booster vaccinations should be given 1 week before major changes in their diet are expected to occur.

Healthcare for Sheep

Keeping good records is the key to a successful plan for your sheep. This in turn will increase flock profitability on an annual basis. Records do not need to be complex or time consuming. The basic requirements are:

- projected lamb crop at scanning and lambs finally sold;
- losses of pregnant ewes, barren ewes (differentiate between true barren and those who lambed but were without a lamb); and
- lamb losses at birth and in the next seven days, losses from seven days of age to end of selling period whilst lambing are always a very hectic period. A simple notebook and pencil to record losses is invaluable in planning for better lambing performances in the future.

Having a good idea of what a healthy sheep looks like and sounds like puts you in the position as a sheep caretaker to identify when something is off quickly.

If you maintain a healthy flock, then you will also maintain healthier profits from year to year.

Sheep Behavior to Notice

Head

A sheep's head should be upright and appear alert. When sheep are well, they look all around. A sick sheep on the other hand, will dip their head downward. The sheep may lay its head down in the corner of the pen or a by a tree if it is out in the field. The only time a sheep's head should be down is when they are grazing. When a sheep is out in the field or in a barn, the sheep should be looking around and noticing what is going on. If you walk in, the sheep should all look at you. If one sheep doesn't, there might be something going on with that sheep. It's worth making a note for further inspection.

Ears

Different sheep have different ear shapes. Any change from the normal shape of your sheep's ears, or the position, might be an indication the animal is ill.

Horns

A sheep with horns is called a horned sheep. A sheep without horns is called polled. Most sheep that are raised for commercial use are polled. Polled sheep are less likely to cause damage to themselves and other sheep. Sheep with horns can get stuck in fences and cause damage to the animal's head. Sometimes a horn can grow around and then poke the sheep in the eye. Do not let this happen. If you have horned sheep, you need to check their horns to make sure of how the horns are growing.

Mouth

The bottom and top jaw of the sheep's mouth should align correctly. The incisor teeth should meet the pad on the upper jaw. You have to look at the jaw from the side of the sheep to check this correctly. Sheep can suffer from overbite and underbite. A severe bite problem can cause a grazing problem. Unfortunately, this might be passed down to offspring. Check the sheep's jaw for swelling in the cheeks. It should not be painful any place in their mouths. The mouth area should also be clean. A healthy sheep does not drool and does not have leftover cud still in their mouth or hanging down from the mouth.

Teeth should also be checked for problems. If a sheep has bad breath, they may be having teeth issues. If a sheep has missing teeth or teeth that are rotting or cause pain, it may have a hard time grazing. When sheep have a hard time grazing, they put on much less weight. When a sheep is older than 4 years, then molar and tooth disease is more common.

Bleating

Healthy sheep will bleat to the other sheep. Bleating is how sheep communicate back and forth. Sheep that don't bleat may be ill.

Breathing

The respiratory rate of sheep is extremely critical. Normal respiration is 12-20 breaths per minute. Breathing should not be labored or followed by a cough. A cough that happens from time to time is ok. If a sheep has been running, they will breathe harder. On a normal basis, they should not have heavy breathing.

Temperature

The normal temperature of a sheep is 38–40 degrees Celsius and 101-104 degrees Fahrenheit. The temperature can vary because of weather, fleece thickness, or if the sheep has exerted itself physically. A sheep's temperature should not be the only indicator of illness. You have to check all of the other factors as well.

Sheep's Pulse

A normal sheep's pulse is between 60–80 beats per minute. If the sheep you are checking knows who you are and is used to being handled, the pulse will be normal if they are well. However, if a sheep is not used to being handled, the pulse rate is not a good indicator of health because the pulse will elevate if they are handled and not used to that act. If your sheep's pulse is elevated and the sheep knows you well, or the pulse is very low, this can be an indicator that something is wrong with your sheep's health.

Form

The form of the sheep is the first thing you will notice. Sheep should have all four feet firmly on the ground and weight should be evenly distributed. The legs should be straight and not bowed or bent.

If the sheep are lifting any legs, or neglecting to place weight on certain legs, this can be an indication of lameness or poor breeding. It is easier to observe lameness when the sheep move around.

Hoofs

When checking feet there are various problems you should watch and check for. These include:

- Abnormal or excessive hoof growth
- Cracked hoofs
- Splayed hoofs
- The presence of pus or other infection

Trim any hoof, if necessary, then have the sheep stand in a foot bath. The foot bath should contain sulphate, formaldehyde solution, and other foot bath chemicals if necessary.

The foot bath is used as a tool for healing, and to further diagnose lame sheep that need feet trimmed. Sheep should come in from the field and go into the foot bath. Unfortunately, the bath will cause any bad hoofs to sting and the sheep will lift their legs or wince. Sheep that obviously have sore hoofs should be separated out and further inspected. The hoofs are trimmed and treated.

Udders

An ewe has to have a working udder if you plan to breed her. If the udder is defective, the lambs will not be able to feed in the way they are supposed to feed off of their mother. Udders should be checked regularly to make sure there are no problems. Look for swollen glands, lumps, or oversized areas on the udders. If anything is found, remove the ewes from your breeding program.

Testicles

A ram's testicles should be examined to see if they have developed normally. Of course, the testicles are important in the breeding program. A healthy ram for breeding should have two evenly sized testicles that should move evenly within the scrotum. They should feel firm and not have any lumps. The testicles should not be swollen in any way.

Sheep Wool and Skin

Sheep skin is hard to see because of all the wool. But if your sheep is losing wool and is not a breed that is supposed to lose wool, then you need to inspect further. Sheep can get some infections such as scrapie, flystrike, and sheep scab. These all need to be treated immediately and will cause wool loss. A sheep's skin should be free from any scabs. If there are any issues, you should check your entire flock to make sure there isn't an infection or a parasite being passed around.

Loss of wool can happen for a variety of reasons. You should look for these other symptoms that might accompany the loss of wool to see if something more serious is going on:

1. Stress. If your sheep are stressed the wool will come off in clumps when your sheep are handled.
2. Illness can cause sheep to lose wool.
3. Itching and wool loss may be an indicator of a parasite.
4. If one sheep is losing wool, it could indicate scrapie.
5. Some sheep breeds lose wool naturally. Know your sheep.

Bottom Backside

Your sheep should always look clean around its backside or rear end. If it doesn't, it could indicate worms, internal parasites, or flystrike. If you see maggots, then you know it is flystrike. Trim the wool, get the maggots off, and then treat immediately with a chemical treatment called Crovect. We don't usually endorse a product, but this does work quickly.

Scoring a Sheep

When you score a sheep, it simply means to assess the form of the sheep. Is this sheep too thin or too heavy?

There is a basic way to use your hand for a quick evaluation: put your hand across the back of the sheep and just over the last rib. A score of 0 means the sheep is very thin (skin and bone), a score of 5 means the sheep is grossly fat. You should consider the maturity of the sheep when scoring. Practice this with your sheep often and also with other sheep breeders. This will help you gain more experience.

Chapter 8: Parasites and How to Treat Your Sheep

Internal Parasites/Worms

Internal parasites, or worms, will cause loss in profit and production within your flock. According to recent studies, internal parasitism is recognized as the most prominent sheep disease farmers see in herds. Sheep infected with parasites may become ill and even die. Infected sheep either don't gain well or lose weight, become lethargic, and may have diarrhea. If losses occur which are undetected because the signs of parasitism are not obvious, it may be because of an internal parasite that is running rampant in your flock.

The internal parasites responsible for the greatest losses to sheep are the ones that infect the true stomach of the sheep. Every flock in some areas in the Eastern section of the United States harbors some of these parasites. The most important of these is a parasite known as the barber pole worm. Another worm called Ostertagia may also infect sheep and cause losses in some cases. A third stomach worm called Trichostrongylus is considered to be less important, but still relevant. Intestinal worms, especially Trichostrongylus colubriformis and Oesophagostomum, may also cause problems. Lung worms are also a problem in some flocks.

The life cycle of the barber pole worm (technical name Haemonchus) is important to learn. When you understand the life cycle, you can effectively treat the sheep or pretreat to beat this ugly parasite.

Adult barber pole worms live in the stomach and lay eggs in huge numbers that are then passed in the manure (see below). Following passage onto the pasture in the manure, they must develop into larvae before they are capable of infecting. The life cycle of this worm might take 5 to 21 days or up to a few months depending on the climate.

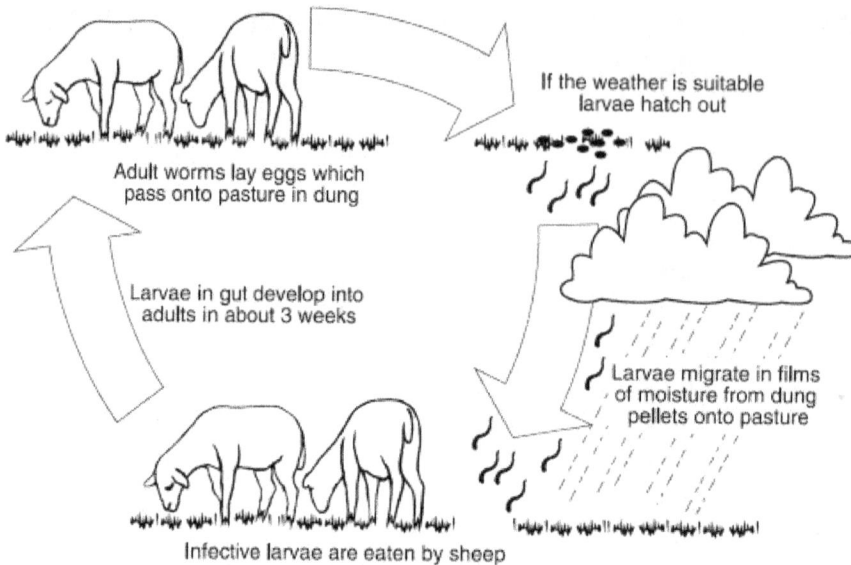

If the weather is suitable larvae hatch out

Adult worms lay eggs which pass onto pasture in dung

Larvae in gut develop into adults in about 3 weeks

Larvae migrate in films of moisture from dung pellets onto pasture

Infective larvae are eaten by sheep

Department of Primary Industries and Regional Development
Agriculture and Food
Government of Western Australia[xxix]

After larvae have developed into the stage where they are infective, they must be eaten by the sheep in order to complete their life cycle. The larvae have a limited ability to transport themselves from the manure onto the pasture plants. Therefore, continuation of the cycle depends on disintegration of manure during rains, which transports larvae in splashes and small currents to the surrounding grasses. As you can see, where sheep graze in dryer climates, this worm is not as prevalent. Little to almost no rainfall will eliminate this parasite. More rainfall equals a higher possibility of barber pole larvae.

When sheep are forced to graze pastures very closely, the number of larvae ingested usually increases because the concentration of larvae is higher in the lower parts of pasture plants. The fact that sheep naturally tend to graze selected areas of the pasture very closely, even when other pasture is available, is one of the characteristics that makes them more susceptible to worms.

What Makes Sheep So Susceptible to Parasites?
Sheep as a group tend to be more susceptible to parasites than other animals. There are several reasons for this fact.

1. The small fecal pellets of sheep disintegrate very easily, thus releasing the larvae onto pastures.

70

2. Barber pole worm is often the major parasite of sheep and its blood sucking characteristic makes it very damaging.
3. The ability and tendency of sheep to graze close to the ground where larvae numbers are higher drastically increases their exposure to parasites.
4. Sheep, unlike many other animals, have very little aversion to grazing areas of high fecal contamination.
5. Sheep have a flocking instinct that encourages them to graze close together.
6. The barber pole worm is a very prolific egg layer, thus worm numbers can build up very rapidly.
7. Even older sheep are unable to develop immunity that controls the parasite life cycle.

Symptoms

Stomach worms cause a number of symptoms. Unfortunately, many of these symptoms do not show up until they are too late to diagnose.

1. The loss of large quantities of blood and protein results in weakness and anemia. If you check gums and eyelids and they are extremely pale, that is one indicator of anemia.
2. When there is a rapid build-up in the number of parasites, sheep may die suddenly due to excessive blood loss, even if they are in good body condition and appear healthy.
3. When the build-up is slower, sheep lose weight, become anemic, and their wool becomes brittle and may fall out. Weak animals may go down, develop pneumonia, and eventually die.
4. A condition known as "bottle jaw" (where fluid accumulates under the skin of the lower jaw) may develop as a symptom of low protein levels.
5. Diarrhea may or may not occur as a result of parasitism. Diarrhea results from intestinal irritation and from disturbed digestion of food. Infections with barber pole worm very rarely result in diarrhea. The other worm species are more likely to cause diarrhea.

It is wise to have a veterinarian inspect the sheep in order to have an accurate diagnosis concerning what is going on. Only after an accurate diagnosis is made can an effective treatment and control program be undertaken. Pasture grass must be inspected, and if sheep have died, an autopsy should be performed to make sure of cause of death.

Parasite Control

Safe Pasture Concept

A very helpful approach to parasite control involves thinking in terms of safe and dangerous pastures. A safe pasture is one where infectivity is low enough that the parasite burdens of susceptible sheep increase slowly. It is not one completely free of infective larvae. Pastures that have been harvested for hay, silage, or small grain crops can generally be considered safe. Pastures that have been grazed by cattle, horses, or other species for a grazing season or longer are considered safe because only a small amount of cross-infection between species occurs.

Contrary to previous practice, a pasture that has not been grazed for a few weeks cannot be considered safe. Now, a year or more without grazing is required for pastures of non-grazing to become safe. Most rotational grazing systems currently practiced do not aid in parasite control. This system usually causes an increased parasite challenge because sheep densities are higher on pastures.

Naturally, a safe pasture should not be used by infected sheep. That's just common sense to any farmer. One control program is based on the concept term "dose and move." The rationale behind this is to extend the effectiveness of a single treatment by moving animals to a safe pasture to limit reinfection. If sheep are treated in early June and moved to a safe pasture, they are unlikely to be exposed to the summer infection in pasture infectivity. If sheep are treated and left on the same pasture, however, they will be exposed to heavy reinfection and derive little benefit from the treatment.

Systems may be utilized where one portion of a pasture is used for grazing during the early part of the grazing season, while the other portion of the pasture is used to grow hay. After the hay is harvested and some regrowth has occurred in early June, sheep may be dewormed, moved to the pasture from which hay has been harvested, and the contaminated pasture is allowed to grow hay during the latter part of the grazing season. The process of drying involved in hay making kills infective larvae on these plants so that this hay may safely be fed to sheep during the winter. Because some build-up often occurs late in the grazing season when dose and move is being practiced, two additional fall (September/October) deworming treatments should be given two weeks apart.

Same Pasture Concept

Since many sheep pastures are not fit for hay production, the rotate and clean system is often not able to be used. In this case, preventive deworming treatments must be administered through the season to provide control of parasites. Such programs must take into consideration the fact that Haemonchus or barber pole worms have a two-week period of development, from ingestion to maturity, before eggs may be passed. Treatment should begin when sheep first begin grazing and are no longer fed harvested feeds in the spring. The use of a product that will kill the larvae for the first treatment is important because it will prevent the development of these larvae into adults that will infect pastures and in turn pass this parasite on to others.

After this initial treatment, several actions can be taken. Sheep may be retreated with an effective dewormer every two weeks for several treatments. However, it is important to realize using treatment at two-week intervals can very quickly lead to the development of drug resistant parasites. Quite the opposite effect of the result you intend.

Such a situation has developed in Australia, where some strains of barber pole worm are resistant to almost every treatment available. The two-week treatment intervals prevent any worms from developing to the point that their eggs are passed in the manure. If a product with a residual effect (that is, a product that persists in the animal and continues to kill incoming larvae) is used, the treatment can be extended by the number of days of the residual effect.

If weather conditions become dry during midsummer, deworming may be discontinued for a time. Remember that pasture larvae levels rebound quickly after a rain, so deworming should be immediately resumed if midsummer or early fall rains come. If you want these programs to be effective, it is essential to include all sheep. Any lambs over a few weeks of age, rams, and replacements must all be dewormed. Leaving even a few untreated sheep mixed with sheep on the program may allow for enough parasite build-up over a period of weeks and months to destroy the entire earlier efforts.

Alternative Deworming Solutions

Some farmers use continuous feeding of a dewormer in the salt or mineral for parasite control. This may provide some parasite control. However, problems may develop because the dewormers available in these forms are not highly effective against all stages life of these parasites. The low-level feeding of these dewormers also encourages the development of parasites that are resistant to the dewormer. The effectiveness will then decrease over time.

When winter lambing is practiced so that young lambs never graze, less strenuous control programs may be needed. This is because all grazers are older and have acquired immunity. Pre-lambing deworming should still be practiced. Remember that young replacements must be grazed separately and given an effective parasite control program. Sheep brought in from arid areas will usually be quite susceptible to parasites and will require an intensive control program. These sheep haven't been as exposed to these parasites previously.

Tapeworms

The tapeworm of sheep (Moniezia) lives in the small intestine and is transmitted to sheep by a small non-parasite mite that lives on a pasture. Sheep are infected when they ingest the infected mites on grass. Although tapeworms are often accused of causing weight loss and/or diarrhea, they rarely cause much damage. One effective drug for treatment against tapeworms in sheep is fenbendazole.

Coccidiosis

Coccidia in sheep occurs when sheep eat feed contaminated with manure, drink dirty water, or graze pastures full of manure as well. These are very common parasites. Most sheep are infected with several different types from an early age. Young lambs are highly susceptible to infection and clinical disease. Older animals may still be infected, but may be more resistant because of their age.

Clinical coccidiosis is seen commonly in young lambs at the time of weaning, in confined conditions, or shortly after entering feedlots; and in sheep which have been physically stressed by weather, handling, and shipping. Sheep on intensive grazing programs may suffer from coccidiosis. Lambs become infected with coccidia by ingesting the coccidia oocysts (eggs).

Through a complicated process, the tiny parasite divides and enters gut cells with more and more damage done to the gut lining. Eventually, the parasite produces new oocysts which pass out in the manure. These oocysts need two to five days exposure to a wet, damp environment before they become infective. If a lamb is infected with a sufficient number of oocysts, the damage to cells in the gut wall may be extensive. This results in watery diarrhea, occasionally containing blood and mucus. Dehydration and weight loss often occur.

If the condition is left untreated, lambs may die. Lambs surviving clinical disease will have their growth potential severely compromised. Coccidiosis can be diagnosed in a live animal by clinical signs and demonstrations of large numbers of oocysts in feces. Sheep in more arid climates do not usually deal with these parasites.

Drugs for Treatment

If a parasite survives deworming, it is said to be drug resistant. This represents a major problem for the sheep industry. Several studies have reported resistance in the major parasite species against several drugs, especially the family of drugs to which thiabendazole belongs.

Four techniques have been suggested for reducing the development of resistance:

1. Use a full dose of dewormer whenever treatment is done.
2. Reduce dosing frequency by decreasing stocking rates or use dose and move.
3. Treat all new introductions with the best products available and perhaps with a double dose.
4. Avoid alternating dewormers during the grazing season. Alternating dewormers between seasons may be advisable in some cases.

The Ivermectin sheep drench product is approved for use in sheep, but does not have as great a residual effect. Ivermectin is a broad-spectrum, safe dewormer that is highly efficient against all worm stages, including hypobiotic larvae. In addition, Ivermectin will provide control of nasal bots and has varying efficacy against external parasites like lice, keds, ticks, and tapeworms.

Treatment for Coccidiosis

Coccidiosis in sheep is usually related to stress, overcrowded conditions, and manure contamination. Frequent cleaning, proper sanitation, and the use of feeders and waterers designed to prevent manure contamination greatly reduces the infection rate and the incidence of clinical disease.

The administration of anticoccidial drugs before anticipated outbreaks can significantly reduce or eliminate clinical outbreaks of coccidiosis in sheep. Products containing the following anticoccidial drugs are commercially available: lasalocid, monensin, decoquinate, and sulfaquinoxaline.

The more you understand about sheep parasites, treatment, and prevention, the better the shepherd you can be for your flock. Parasites are most likely going to show themselves if you live where there is a rainy season because parasites are able to grow quickly in the fields and in feed. However, with pretreatment and effective parasite control programs, you will be able to raise your flock with the least amount of parasite problems possible.

Chapter 9: Shearing Sheep

There are two ways to shear sheep. The first is to shear sheep yourself and the second is to hire professionals to get the job done. If you are going to shear your sheep, you should first observe and then learn the steps before attempting it yourself. We briefly discussed this in Chapter 5 under Shepherding Skills. Now, it is time to discuss shearing sheep in detail. All sheep should be shorn in the spring. Even if you are not selling the wool for profit, the following reasons explain why it is critical to shear every sheep:

1. Wool will keep growing if not shorn. This will make your sheep uncomfortable as the months get warmer. As the wool gets heavier, the sheep will pull on the wool and the wool may come off in clumps, causing sores on the sheep's skin. The sores will attract flies that may infect the skin and then cause flystrike.
2. If your sheep do not get sheared, they may get itchy. If this happens, they could possibly rub on hay racks, each other, and fences.
3. Flystrike. We cannot talk about flystrike enough. This problem if left untreated will be fatal for sheep. Flystrike is caused by flies laying eggs in sore areas in thick, matted wool usually near the buttocks and underbelly. When the maggots hatch, they literally feed on the sheep's flesh. The idea behind shearing sheep is to get rid of the dirty wool before any fly's eggs can hatch.
4. Early shearing gives the shepherd more chances to assess the sheep's condition after the winter. By the condition, we are referring to the health, weight, checking the eyes, ears, mass of the body, hoofs, and of course, the tail area where flystrike always seems to happen most often. Check for redness in the genital and urethra area. Also, look for any irritation or itching in those areas.

The first time you shear your sheep, hire a professional and observe what they do. This is how you learn to shear sheep yourself. If possible, shear a sheep or two with the trained professional right beside you to guide you while you work. This job takes muscle and is physically draining. You are going to need certain supplies to shear your sheep. Whether you are shearing or you are hiring this out, you should buy the equipment so you have it on hand. Below are the necessary supplies:

- Large tarp to cover the ground and catch the fleece as it falls
- Plastic bags for temporary storage of fleece
- Treats for your sheep
- Water for you or the people you hire to drink
- Broom for sweeping between animals
- Extension cord for the shearing clippers
- Blue-Kote, Swat, Cornstarch in case there are accidental cuts or scratches to your sheep
- Scissors for cleaning up around ears, tail, and other areas you cannot get with shears
- Shearing clippers for sheep shearing

The shearing clippers are quite expensive. It is important to use them properly and take care of them while shearing and after the shearing day is over. The clippers are used with a shearing blade and a shearing comb. Make sure you have the proper comb for the fleece of your sheep.

The fleece should be sheared in one piece. People who are used to shearing call it unzipping the sheep. The chest area or brisket of the sheep is sheared first. Keep yourself close to the body to avoid doing the same area twice. This is when having sheep trained to harness comes in very handy. If you go over the same area twice, then the value of the fleece decreases. Spinners cannot deal with the short second cuts.

After you shear the chest and lower neck, then you are going to work on one side of the sheep and after, part of the back to the other side. This process is repeated on the other side and then the rest of the back. The whole piece will then be placed in a bag. After you have the entire fleece, you will then shear the legs, the dock area, and the crotch. The fleece from these areas are usually swept up and discarded. If you can learn to shear your own sheep, this will save you money and increase your wool profit.

Chapter 10: Reproduction and Lambing

If you are raising sheep for profit, then your most single important factor in raising sheep is production rate of your flock. Productivity of the ewe group is a direct reflection of the reproductive efficiency and in turn your profit. You may have the highest quality sheep with a background of genetic merit, beautiful eye appeal, and great placing. But if your sheep does not reproduce, you might as well keep it as a pet. It is not going to help you reach your goals.

If you are proactive, you probably have goals and objectives for your next lamb crop. Usually these are fixed before and during the breeding season. Increasing ewe productivity while decreasing labor, time, and facilities used for these purposes during the lambing season might be some of your realistic objectives.

Reproduction in sheep is generally influenced by many factors. These include genetic background, potential, nutritional status, the environment you created, day length, health status, and the overall happiness of your flock. These factors are critical in both the ewe and the ram. Many sheep have potential for multiple births. We can use management practices to influence these factors to increase the reproductive rate.

Even though sheep may be a business, they are still living things. You should spend as much time as possible among your sheep. Each sheep should know you and should be glad to see you. Their lives should be happy while they are in your care. Your production rate will be higher if you have a more secure and happy flock.

Reproduction in the Ewe

Most breeds of sheep have a seasonal breeding pattern. There are some breeds that have two breeding seasons. Ewes that have only one breeding season will begin to cycle when the daylight begins to shorten.

The decrease from light to dark triggers hormone changes. This period begins in late August and extends through January. The peak fertility time is September to October. You do not want lambs born in the dead of winter, but it happens more often than not. This requires more in the care for the lamb in terms of heat and protection. Ewes cycle every 16–17 days until they are bred. If fertilization occurs, they will give birth between 144 to 152 days after they mated.

Reproduction in the Ram

A ram that has been taken care of can settle more than 75 ewes during breeding season. If your ram is unhealthy, he will settle zero. The moral of this story is to take care of your rams and for that matter, take care of all your sheep.

To determine fertility capacity in the rams, a breeding exam can be conducted. This includes visual appraisal of the overall health and wellness of your rams, and a check on their feet, legs, eyes, mouth, teeth, and basically their entire body. You need to check their testicle mass and the scrotal circumference. Testicle mass should be firm but with no abscesses, injuries, or any other conditions that would affect fertility. The penis should be examined to be determined if there are indications of injury or adhesions. When measuring scrotal circumference, it is important that both testicles are fully descended. Always take the measurement at the point of the greatest circumference.

When checking both ewes and rams, this is where harness training comes in handy. This is also where having a good relationship with your flock pays off. If your sheep are harness trained, these checks are much easier. You can put on the harness, lead the sheep to a place where you can hook the ewe or ram to a fence, and then talk to the sheep and do your check. It is much easier to do if your ewe or ram is used to you and the harness.

Temperature of Environment

Sheep are less likely to participate in reproductive activity when the temperatures reach above 90 degrees Fahrenheit and extreme humidity. In truth, it is more the rise in body temperature than that of the environment. The importance of shearing in the spring cannot be overstated. High body temperatures can occur from heat and also from stress. Again, the importance of a happy flock is a way to more profits as well.

Fescue pastures often grow a fungus. If that happens, when the sheep ingest this fungus, their body temps will also go up and reproduction activities will go down. Ovulation rate slows when the body temperature is higher. And even more important, rams that are heat stressed can be sterile for 6 to 10 weeks.

Some common sense measures can be taken to help your sheep to prevent this heat stress. Sheep can be sheared again right before breeding season. Provide tons of shade during the heat of the day. Try not to add stress to your flock by moving them during the heat of the day.

Nutrition

Two to four weeks before breeding season, ewes can be flushed. This means increasing the quantity of grain on top of their grazing so they gain more weight. This will usually increase their lambing rate. The flock should also have access to salt and mineral mix without the copper toxicity problems. If ewes are healthy and fit it is more likely they will have multiple births and will settle on their first service from a healthy ram.

In Summary

Reproduction in ewes and rams is something you as the farmer and shepherd have a lot of control over. As you have read, nutrition, careful checks on your sheep, stress of the sheep, and other factors are going to decide the production rate of your flock. The critical point is to know and understand your flock and for them to know you. The stress factor will go down considerably if the flock knows the farmer when you are the person who runs inspections and do the shearing. If you do not do the shearing, you should be on hand so they see you. The trust you build with your flock will reap profits later.

Lambing

We discussed lambing in Chapter 6, The Business of Raising Sheep for Meat. But we should revisit lambing again because not every flock is being raised to sell for meat.

Lambing is the most important activity in your flock. Let's say for argument's sake you raise your flock for wool. Your success during the lambing season is going to ensure a successful shearing season the next year. More lambs equal more wool.

To Prepare

As lambing season gets closer, the ewes that are going to reproduce should be sorted out as they will need more attention. This pen of particular ewes should be locked up at night. Do not put them in lambing pens. It's not time for that. Just put them in one of your places to be separated.

Here is a list of things to do before lambing:

1. Increase level of ewe nutrition six weeks prior to lambing

2. Crutch or shear ewes
3. Booster immunizations which may include C, D, and T
4. Trim feet
5. Clean out barn and set up lambing pens
6. Check heat lamps, feeders, and water buckets
7. Get lambing equipment supplies together

Lambing Equipment Supplies Checklist:

1. 7% Iodine
2. Towels
3. Plastic sleeves and disposable gloves
4. O.B. lube
5. Scissors
6. Ewe restrainer
7. Baby bottle
8. Stomach tube and two-ounce syringe
9. Uterine boluses
10. Halter
11. Lamb milk replacer and colostrum replacer (if needed)

When the ewe is about to give birth, the udder becomes engorged, swollen, and slightly red. Ewe lambing signs include the vulva stretching out and becoming red and swollen. Often an ewe will miss a feeding or separate herself from the flock before labor begins.

Sheep Birthing Process

The sheep birthing process starts when the ewe begins contractions and begins getting up and down frequently. She will paw at the ground. The ewe will lie down and push with her nose up in the air. When this is going on, leave her alone and do not disturb. Observe her quietly from a little distance. If you have a very close relationship with this ewe, you maybe could talk to her with your soothing voice, but most likely, she wants to be left alone.

The water bag will come out first. After the water bag, the lambs will be delivered within ½ hour to 1 hour. Make sure the lamb's mouth and nose are clear of mucous and fluid. Make sure the lamb is breathing. The ewe will claim the lamb. She will then dry it off. You can tell if she is going to have another lamb. If she is not, put the ewe and lamb in the lambing pens if they are not already there. If the ewe will let you help with the lamb cleaning go ahead. If you have a good relationship, she will be pretty proud and want you involved. Put a little straw in there. Set up the heat lamp. Make sure the lamp is higher, so it doesn't burn the lamb. Clip the navel of the lamb to 1 inch long and then dip it in iodine.

Make sure milk is flowing from the ewe's teats. Check to see that the lamb or lambs are latching, then give the ewe a flake of alfalfa and water with the bottle. Give the ewe lots of love. She deserves it. If all looks well, you can leave and check again in an hour. When you return, make sure the ewe is still owning the lamb(s) and nursing or the lamb(s) are sleeping by the ewe. Give the ewe another alfalfa treat and water. She will push the afterbirth out on her own.

How to Help an Ewe Give Birth
If an ewe is having problems giving birth, you have to help get the lamb out. Use the ewe restrainer to contain the ewe to help keep her from further complicating the matter. Use the sleeves and O.B. lube that you have with you. Check the ewe to make sure she is fully dilated. Make sure what you are looking at as far as the legs and head belong to the same lamb. Alternate pulling one leg and then the other at a slightly downward angle until the lamb comes out. Place the lamb in front of the ewe.

Handling Lamb Death
Sometimes lambs die no matter how hard you try to save them during the birthing process. The ewe does all she can and so do you and yet a lamb may be born dead. If an ewe has a dead lamb, dispose of it quickly. This includes the birthing fluids, etc. Use disposable gloves when handling everything. If the ewe only had dead lambs, move her into a lambing pen that has been strawed. Decrease the quality of hay and provide no grain supplement. Give her water. The idea is to reduce energy so the ewe can rest and recover without mastitis complications. You want her udder to be saved and for her to dry up. In some cases, you may be able to pair her with another lamb.

Good Advice

- During lambing season, check the sheep barn and feed those sheep first.
- Do not wear heavy perfume or cologne around a newborn lamb or the ewe as this can cause the ewe to reject her lambs.
- Be very careful when handling lambs from more than one ewe. Wash hands in between if possible or use disposable gloves. Mixing the smell of one newborn to another can cause an ewe to reject her own lamb.

- To get a newborn lamb breathing quickly, you need to stick a small piece of straw up its nose. This will make the lamb sneeze. Make sure to wipe away excess mucus or membranes first. You can also rub the lamb's ribcage to get the lamb moving better.
- Getting colostrum into the newborn lamb in the first 15 minutes is critical. This warms up the lamb, gives it energy, and will supply the necessary antibodies.
- When attempting to get the newborn lamb to nurse, tickle the lamb under the tail. For some reason, this stimulates suckling.
- When you do carry the newborn lamb, keep it as close to the mother as you can. Carry the lamb low so it can be near her. The ewe will follow the lamb into the lambing pen.
- When putting iodine onto the lamb's navel, tip the lamb up with the bottle. Iodine will stain your clothes. When lambing always wear farm clothes.
- Make sure lambing pens are always clean.
- Don't interfere if you don't have to.

Chapter 11: Marketing Sheep

Marketing Sheep and Wool

All sheep operations need to market products in order to generate income. To be successful and prosperous in your sheep enterprise, you should understand basic marketing concepts for selling sheep and their products. This includes specific business models of marketing lamb for the holidays, marketing lambs at other times of the year, marketing breeding stock, and marketing wool and woolen products. As a sheep producer, what sort of plan do you have for marketing your lambs?

One of the first steps to consider when starting any business is to develop a detailed marketing plan. This plan will serve as a blueprint for your business and covers everything from how you will go about targeting your customers, to calculating profitable prices.

Simply put, a marketing plan lays out your strategy to achieve your business goals and objectives. More precisely, it is a clear and detailed "map" for how you will sell your lambs or their products to your customer. Your marketing plan specifies how you will handle certain tasks such as: getting your lambs to market, building a market niche, identifying your customer base, maintaining sales growth, or dealing with future changes to your operation.

Your plan also serves as a useful means of accumulating valuable data and gauging day-to-day tasks and events. For instance, you might use your plan as a template for marketing tasks that occur throughout the year. In addition, the marketing plan might note when you need to attend to specific activities, such as registering sheep, sending performance data to the National Sheep Improvement Program (NSIP), or setting up a processing schedule so that you have adequate inventory of lamb for festival markets.

To prepare your marketing plan, you will need to start with a basic concept for the plan: position.

Positioning involves influencing how your customers perceive your business and your product. It creates your unique farm identity and puts your lambs or lamb products in people's minds. In turn, when consumers are interested in buying sheep or lamb products, they will think specifically of your farm or unique brand.

A **brand** or **branding** is the marketing practice of creating a name, symbol, or design that identifies and differentiates a product from other products. For this reason, your brand is important—it is what distinguishes your products or services from those of your competitors. It also helps keep your customers focused on your image and helps maintain and grow your sales.

As a sheep producer, there are some strategies that you can employ that will help ensure good positioning and branding, or that "customer allure," to entice sales. For instance, you could:

- Develop a farm name and logo that serves to identify your product or products
- Sell pasture-raised, or grass-fed, lamb for a unique taste and quality
- Offer organic lamb
- Highlight your production practices as a selling point
- Identify genetic and/or visual characteristics that appeal to buyers interested in breeding stock
- Highlight wool characteristics that appeal to hand spinners

Identify the unique aspects that differentiate your product from other sheep or lamb products. For instance, you might want to note the genetic merits of your sheep through NSIP to identify outstanding estimated breeding values (EBVs) in order to sell breeding stock. Alternatively, if you elect to market lamb, you could use any unique attributes to distinguish your product from your competitors' lamb products. Remember, **your buyers want to feel confident that your product is superior and consistent,** and you will have an excellent market position.

Spending a few minutes developing a marketing plan can keep you on track to develop sheep that best meet your customers' needs and that also best use resources available on your farm. By keeping your standards high and offering a quality product, you're sure to have loyal buyers who will spread the word about your farm and the products you offer for sale.

Marketing Sheep for Meat

Lamb production has shown positive growth over the last decade. Strong consumer markets for lamb have meant an increase in both new and experienced producers who want to get into commercial lamb production. Lamb production remains an attractive lifestyle and small business for many with a limited land base. Whether you are a small flock owner or have a large flock business, the success and reputation of the industry depends on your producing and marketing high-quality, safe, and delicious lamb.

Producers are, however, in control of many aspects of production that can positively affect their returns on market lambs. A branded product, as mentioned above, with specific attributes verified through new tools such as traceability, welfare, or halal certification may command a higher price.

Moving away from commodity status into a more specialized marketplace often improves returns. The potential in specialized, or niche, lamb markets is an opportunity for the lamb supply chain. For many producers, raising meat lambs is the focus of their operation. Wool and milk production are very small markets and will not be the focus of this publication. Many lamb producers struggle for positive returns, even when prices are good. Very broadly, business success in lamb production depends on three factors:

- How the business is managed: managing such factors as cost of production, flock productivity and lamb quality is critical.
- How the lamb is raised: management practices enable the producer to provide the quality and style of lamb the market wants.
- How the lamb is marketed.

It is crucial that lamb producers have a marketing plan established before they begin raising sheep. Knowing where you will ultimately market your lambs impacts your whole production system. For example, if you are selling direct to consumers and they are looking for grass-fed lambs, you'll need a breed type and management system that supports that. If you plan to finish your lambs yourself or through a feedlot, which is not advised in this book, your production system may include feeding grain to your lambs early, even if they will be reared on ewes on pasture.

Again, when planning your production and marketing systems you must focus on where you will market your lambs. The market option you target will also impact the breed type you choose. A breed type can determine traits such as strong mothering, milking and reproductive traits, or terminal sire type where all lambs are marketed for meat. A breed type can also be wool or milk producing. Within each breed type are numerous breeds of pure genetic seedstock.

Some breeds—like Suffolk—are terminal sire breeds. Suffolk rams are used when the primary objective is to produce good-quality market lambs. Suffolks can provide the genetics to boost lamb growth, feed efficiency, and carcass quality. Their offspring are strong contenders for feedlot finishing.

Rambouillet is a range-type, maternal breed type selected for wool production and their ability to thrive under harsh conditions. Using a terminal sire breed on a Rambouillet ewe will improve the carcass quality of the lambs. Using a maternal breed—Rideau, Dorset—on range-type ewes can also improve lambing percentage and milk production.

Planning ahead for your market, then, should be a key factor that drives the selection of the breed or breed type you choose to raise. Lamb producers have a number of market options available to them. This gives producers the chance to build a business and operation based on the best market fit for their lambs. Whatever market option you choose, your lambs must have an ear tag before they leave your farm, the farm of origin. Transporters, lamb buyers, and/or auctions and processors are prohibited from accepting sheep and lambs unless they bear an individual tag in almost every jurisdiction.

Lamb market options:

1. A federally inspected plant. This is the main market for approximately 70% of the lambs produced.
2. Lamb feedlots. Currently there are several feedlots that buy and/or custom feed 30,000 to 50,000 lambs per year in total.
3. Lamb buyers and dealers. There are a number of bonded and insured lamb buyers/dealers who operate.
4. Auctions. A number of auction businesses handle sheep; some have seasonal sheep sales, while a few hold regular sheep sales. Breeding stock is not generally purchased through auctions, where the disease transmission risk is higher due to the co-mingling of multiple flocks.
5. Provincially inspected plants. Provincial plants either purchase and slaughter lambs for their own retail sales outlet, or custom slaughter for producers.

6. Producer or farm-direct marketing to consumers, through the freezer trade, retailers or restaurants, and food service. Between 20,000 to 25,000 lambs per year move through provincially inspected plants. Niche and specialty market options that cater to special consumer groups include lamb certified halal, kosher, or organic. Certification complies with national regulations for labelling a product. Other specialty lamb might be based on a production system (e.g., grass reared and finished), or on a specific type of sheep (e.g., Icelandic lamb).

There are tools available to help producers validate specific traits. Part of a marketing plan might be the use of the On-Farm Food Safety program, an animal welfare program and traceability. What is the best way to market your lambs? The fact is, there is no single approach or market option that works for all producers, all the time. Sometimes no market works well!

A balanced marketing plan includes more than one market option. The purpose of this module—Marketing Your Lambs—is to outline the marketing choices available to you as a lamb producer and to explore the opportunities and demands associated with each. The first part of the module will introduce lamb market opportunities in more detail to show the drivers of lamb production and lamb meat marketing.

Lamb market prices—as with any commodity—fluctuate by nature, depending on supply and demand. There are no regular cycles in lamb market prices, but despite significant price volatility, there is an upward trend over much of the last 15 years. Prices are, however, currently lower than the record highs in 2011. Lamb has also seen an increase in consumption in Canada and North America, while other types of meat consumption have softened in recent years.

Lambs can be marketed at a couple of months of age (Easter lambs), but some are barely weened, and up to one year of age—through a variety of market options—so lamb producers have some flexibility in timing their marketing. To be marketed as lamb, the animal must be less than one year of age; this is verified through dentition (fewer than two permanent incisors) and the characteristics of the break joint after slaughter. Animals over 12 months of age are classified as 'mutton' and are significantly downgraded in value.

In 2012, China surpassed France as the world's largest sheep meat importer. Decreasing farmland, increasing population, and urbanization in China meant internal or domestic sheep production could not keep up with the demand.

China is primarily interested in mutton and secondary cuts (caps and flaps traditionally used for pet food in North America). This meat can be sliced thinly and is used in fast food 'hot pot' restaurants. In the past 10 years, lamb imports into China (primarily from Australia and New Zealand) have increased 500%.

Feedlots

If you are new to sheep farming you may not be aware as to what a feedlot does after you sell your sheep. Keep in mind if you are selling your sheep for meat, it is a rough business, but a feedlot may be a very big shock to the system of a lamb who is used to grazing in an open field. A feedlot is an animal feeding operation that uses corrals, confined pens, or enclosures to grow animals until they are ready to market and slaughter for meat.

'Ready to market' means that the lambs are the correct weight and have the right amount of fat finish required by a specific target market. Feedlots are used to increase efficiencies by feeding larger numbers of animals, to reduce finishing costs per animal, and to produce more uniform carcasses. Feedlots either buy lambs from producers, through auctions, or from livestock dealers. Lamb feeders may also custom feed lambs for a set rate based on their feed, facility, labor, handling, and shipping costs. Some feedlots want the producer to deliver the lambs, while others may arrange transport, which usually involves an added charge.

For small flock owners or producers without a way of transporting lambs, this can be an attractive option. Transportation can be a significant cost, particularly for producers with a small number of lambs. Delivery arrangements should be part of the contract negotiation.

Upon arrival at the feedlot, lambs are identified (if necessary), weighed, and sorted into uniform groups. Feedlots use tags applied on the source farm or on arrival at the feedlot to electronically track and manage lamb nutrition, growth, and healthcare. Lamb health is carefully monitored in the adjustment period as they are introduced to new rations.

Depending on the weight and age of the lambs, the target market, and planned slaughter date, the lambs may be fed a 'grower ration' or a 'finishing ration.' A grower ration allows for a slower, longer growth period while a finishing ration provides for more rapid growth and a pre-determined fat finish. Many lambs are finished on a higher-energy grain-based diet (usually barley) until they reach their ideal finish and weight. Grain finishing improves carcass quality and consistency and contributes to the eating quality preferred by North American lamb consumers. Lambs are efficient converters of feed to meat and are able to utilize grains of a type or quality unsuitable for human consumption.

Photo by Malin Strandvall on Unsplash

Electronic systems have provided lamb feedlots with the ability to efficiently handle and sort lambs of different types and weights into different feeding pens. Feedlots visually monitor and weigh each group of animals to ensure as many lambs as possible are healthy and growing to meet optimum finish for the market they are selling to. When lambs reach the finished weight, larger groups are available for shipping to a specific market or processor. As a result, lamb feedlots can custom produce for different target markets, allowing both the feedlot and the producer/supplier to maximize returns on lambs.

There are various agreements that can be made with the feedlot operator with regard to the purchase and sale of lambs. Producers can sell their feeder lambs outright to the feedlot and be paid on the delivery date, and the lambs become the feeder's property. Producers who want assistance with finishing and marketing lambs have the option of retaining shared ownership and having the feedlot provide custom feeding. In this case, producers pay the feedlot to finish the lamb (for a set cost per day, or for cost of feed plus a set fee), and the producer then selects the market to sell into.

Disadvantages to Selling to Feedlots

- If your animals have only been grass fed, the grain-based diet at the feedlot may require a period of adjustment for your lambs, or you may have to slowly start your lambs on grain.
- If your lambs are not healthy or robust, the feedlot may discount the price they pay—or may reject the lambs—since unthrifty lambs can have problems competing and are prone to injury or death in a feedlot environment.
- If feedlots are not located near your farm, long-distance trucking may increase your costs.

What Are Feedlots Looking for in Lambs?

Feedlots want healthy, consistent, clean lambs. They are looking for lean, fast-growing lambs that are healthy and compete well in a feedlot environment. Many feedlots are also interested in a year-round supply of lambs—your lambs may fit one of those time periods when they are short of lambs.

Most feedlots buy lambs over the phone, so a good deal of trust is involved. To build a great relationship with your feedlot, know and communicate your animals' condition accurately. Deliver or sell what you say you will. You must know the weight of your lambs. A scale is an essential tool for today's lamb operations.

Lambs should be healthy and eating well. Very thin or 'poor-doing' lambs will likely be discounted. Before you call the feedlot buyer, know the weights and body condition of the lambs. Be ready to tell the buyer what the lambs are being fed and if there are any problems (e.g., excess manure tags). If animals arrive at the feedlot and don't meet the weight or type discussed by phone, the price will likely be adjusted. It takes time and trust to develop agreements that meet the needs of both parties. Ensure you establish precise terms for your transaction and be sure you deliver on those terms.

You can do a lot to prepare your lambs for the feedlot environment. Ensure they are strong, have good energy, and also have enough weight on them to compete with other animals. Typically, a feedlot is looking for bigger, highly muscled, lean lambs with good cover. Some feedlots will pay a premium for lambs that are in particularly excellent condition or that have been started on grain.

Prepare lambs for weaning by providing access to a creep feeder with a ration containing whole grains, supplements, or pellets.

Weaning occurs when winter-born lambs are between 45 to 60 days of age. In spring lambing systems where the ewes and lambs go onto pasture, weaning occurs when lambs reach about 80 lb. or when pasture quality is inadequate to maintain good lamb growth. Lambs on pasture with their mothers should have exposure to some grain so they can be weaned with less stress onto dry rations, and eventually adapt more readily to feedlot conditions. Lambs weighing at least 80 lb. tend to compete better in the feedlot than smaller, younger lambs.

Healthy lambs have a better chance at successful transition to the feedlot. Weaning, handling, sorting, and trucking are stressful and can result in weight loss, illness, and death. Build flock immunity through a regular vaccination program for both the ewe flock and lambs. Have a general flock health program where you monitor the health of your ewes and lambs, isolate and treat sick animals appropriately, and keep accurate records of all health treatments. Be prepared to provide this information to all buyers.

Be sure to ask the feedlot what they are specifically looking for. Depending on the market they are selling into, they will look for different types of lambs or attributes. Invest in a weigh scale for your farm so you can manage lamb growth and maximize returns by providing lambs in the weight range the feedlot wants. Generally, feedlots are looking for lambs around 80 lb. and do not want to buy lambs under 60 lb. or over 105 lb. Feedlots like to purchase lambs that have been weaned, are on feed, and that are roughly four months of age.

The goal of the feedlot is to finish lambs that meet their market requirements. Feedlots will pay more for breed types they know grow more rapidly and are more efficient in converting feed to gain. Breed types that grow slowly, don't compete well in a feedlot situation, or are discounted for hide types will likely be priced lower to compensate for additional feedlot costs. Feedlots can provide a quote for lambs based on the current market price. However, when market prices are fluctuating and lamb supplies changing from day to day, they may only hold that price for a short period.

Feedlots are in business to make money. Don't expect any business owner to waste time while you ponder your options. Building a relationship with a feedlot means being considerate about the feedlot buyer's time. The lamb market is competitive. Feedlots buy lambs from many producers, from other provinces and—when the exchange rate is favorable—in the United States. Feedlots favor producers they have success with, who provide consistent quality lambs and deliver what they promised. If selling lambs to a feedlot is part of your business marketing plan, make sure you are a preferred seller.

Transportation to Feedlots

Careful handling of animals during loading, unloading, and transport is critical. The investment you make in a beautifully raised and finished lamb can be undone in the last days of its life through poor handling and transportation. Stress results in weight loss and can contribute to health issues and death. Rough handling can cause carcass damage and bruising, resulting in higher trim losses at the plant.

- Be sure to ask if the feedlot wants lambs off their feed prior to trucking, and if so, for how long.
- Plan for the price impact of weight loss (shrink) during transport to the feedlot from shipping stress and being off feed and water.
- Other factors impact shrink: shrink is highest in the first 35 miles or 40 kilometers.
- The longer the time in transport, the greater the shrink.
- Lambs lose more weight in hot weather than in cold weather.
- Lambs consuming grass or forage diets will shrink more than those consuming concentrate diets.
- Young lambs shrink more than older lambs; five- to eight-month-old lambs usually shrink 5% or more during transportation.

Reconnect with Feedlot after the Sale

Producers can learn a lot from experienced feedlot owners and managers. While their time is valuable, they will see benefits from building a business relationship with their lamb suppliers. If you are interested in improving the sale or experience the next time around, take the time to touch base.

For example, if you have a shared ownership or custom feeding arrangement, ask the feedlot to provide information from their tracking systems on how your animals gained and how well your lambs performed in the feedlot environment. Talk to your feedlot about recommendations on breed types that do well in the feedlot. For example, Suffolks (or meat breed crosses) are known to grow rapidly and thrive in confinement. This is an example of goodwill that will move you to the top of the list as far as producers go on the feedlot's list.

Auctions

Auctions sell livestock on consignment to buyers through an open bidding process. Livestock is sold in a ring at a prescheduled time where hopefully many buyers come to bid on the animals. The animals are sold to the highest bidder. Some auction sales will present graded or sorted lambs in larger lots (or in lots sorted by gender or type), while other sales will offer each producer's lambs separately.

Auction businesses hold regularly scheduled livestock auctions and advertise or post the upcoming sales on their websites. To be compensated for its part in facilitating the livestock sale, the auction business will receive a commission for the sale of the animal and may also charge yardage fees that support their infrastructure costs.

Auctions draw buyers from a network the auction business has fostered over time. The number of buyers attending a specific auction can vary, depending on the auction's lamb network, customer base, time of year, and/or market demand. While producers can choose to sell at any auction, the distance from their operation to the auction should be a consideration because of the cost of transportation and the impact of shrink on the lambs.

Whatever auction you choose, a strong relationship supports their business needs as well as your own. Producers who want to sell their lambs at auction can schedule them into a regular or special sale. Auction businesses may have sales occurring every week, every two weeks, monthly, or at certain times of the year.

Advantages of Selling at an Auction

Depending on the producer's situation, auctions can be a suitable option. For example:

- It is a convenient way to match buyers and sellers together in one location.
- Public livestock auctions provide price discovery, where a price is arrived at through the interaction of buyers and sellers.
- Selling is available year-round and sales can occur regularly.
- If you have a small flock, auctions can sort and group your animals into larger lots of lambs to appeal to a broader buying audience.
- Auctions take a variety of animals: light or finished lambs, cull ewes and rams. Often there are several buyers bidding against each other on every class of sheep or lamb (e.g., light lambs can go to buyers seeking that weight or to be sold as feeders).

- Auctions have their own network of buyers from finishers to processors, so lambs are available to buyers looking for a variety of weights, traits, and breeds.
- Auctions want successful sales so they do a lot of the marketing for you to make sure there are buyers and competitive pricing for your lambs.
- Payment is guaranteed and prompt, and in most cases is processed within an hour of the sale.

Disadvantages of Selling at an Auction Might Include:

- The price you receive for your lambs will be a total unknown.
- If there is a poor turn-out of buyers, prices will be poor.
- Depending on the time of year you sell, you may be subject to wide fluctuations in the market as local supply and demand changes.
- There may be added fees to pay when compared to other forms of selling (sales commission, yardage, and insurance).
- The market prices paid at low-volume auctions may be less than the market prices paid at high-volume markets.
- Lambs marketed at auctions can be subject to increased stress due to noise and handling at the facility.
- If the auction is located at a significant distance from your farm, shrink and transportation costs can lower your returns.
- If you set out to produce the highest quality lambs, selling your lamb at auction may not bring the differentiated returns you seek.
- Biosecurity protocols are needed when feeder lambs are purchased from auctions.
- Because of the numbers of sheep and lambs that are mingled and pass through an auction, the risk of disease transmission is high and makes purchasing breeding stock riskier.

What Are Auctions Looking For?

- Auctions are looking for clean and healthy animals in a variety of sizes.
- Auctions value a consistent supply of lambs year-round.
- The time of year can impact the type of lamb the auction wants, so connect with the auction early in your production cycle to find out what they are looking for and when they need them.
- Auctions prefer advance booking as that gives them time to find the right buyers and advertise your lambs before the auction.

- If you have a sick, weak, or otherwise compromised sheep or lamb, do not bring it to the auction; it is not only inhumane, but presenting poor-quality lambs at auction tarnishes the reputation of your farm and of the auction business.

Check out auctions in your area before you decide to sell. Talk to other sheep farmers and see how the auctions usually work. Do the sheep sell high? Low? Attend a few auctions whether you are selling or not. Make sure you check your total marketing cost of an auction vs. a feed lot vs. private selling. Always know current market prices before talking to anyone about selling your animals.

Marketing Wool

Sheep Types and Their Wool
The Merino sheep and its crossbreeds are the basis of southern hemisphere fine wool production. The Merino originated in Spain. The breed grows well in arid conditions as found in Australia, South Africa, and parts of New Zealand.

The Merino of Australia is the backbone of the largest wool producing country in the world and this breed is the only one grown purely for its wool. The Merino ranges in micron from superfine (12–13 microns) to coarse (25–26 microns), the finest grown in Australia. The bulk of Merino wool production is 20–23 microns. Staple length varies from 30–90 mm. This breed is found in many countries of the world and the quality of fleece produced varies greatly, depending on growing conditions and animal husbandry.

Norwegian
There is more than one breed of sheep in Norway. The oldest is a luster wool breed known as Gammel Norsk Spelsau (translated as Old Norwegian Spelsau). The Spelsau is a breed with coarse outer hair and shorter, finer second growth. It is related to the Gotland and also the breeds of sheep found in Iceland and Faeroe.

However, the main breed in Norway is now a crossbred sheep produced by crossing the Cheviot, which was imported from the UK in the 1800s, and the Dala and Steigar breeds, native to Norway. The wool produced from the first clip is shorn in summer. It is approximately 29–36 microns and 80–120 mm long, which makes it suitable for combing. This wool is suitable for felting, hand knitting yarns, and woven garments, where good resilience is required.

Shetland

The Shetland is the smallest of the British breeds found mainly on the Shetland Islands. It is believed to be of Scandinavian origin. The breed produces wool in several shades, including white, brown, grey, and black. The wool is fine, soft and silky to the touch with a good, bulky down characteristic.

Production is fairly small and much of the clip is consumed by the islanders themselves. The wool varies in quality from approximately 28–33 microns and fiber length from 50–120 mm. The name 'Shetland' has become generic. Much of the knitwear available in the general marketplace is not produced from Shetland wool at all, but from wools of other origins, which have a similar quality and appearance.

The origin of the Jacob or Spanish sheep is not known with any certainty. The first flocks in the UK were based on stock imported from the former Cape Colony, having been established there by settlers from Spain and Portugal. The fleece is mottled/patchy in appearance with the dark patches becoming lighter as the sheep matures. This breed is in demand for handmade textiles as the range of colors produced are more varied than other breeds. The quality of the fiber ranges from approximately 32–40 microns and a length of 80–150 mm.

Blue Faced Leicester

Blue-faced Leicester wool is classed as long-wool with luster. The breed evolved during the 19th century and originally came from the Tyne and Wear valleys and hills of east Cumbria. It was sometimes referred to as 'Hexham Leicester.'

The wool is fine and dense with a good luster, and is long. Therefore, it is well suited to combing. The sheep produces a fairly small weight of fleece for its size and the fleece has been highly prized in recent years for its likeness to mohair, for production of attractive lustrous yarns with good resilience. The fleece is available in white and natural brown hue.

Masham

Masham, pronounced *massam*, is a cross of Teeswater or Wensleydale ram with Dalesbred or Swaledale ewes. The fleece is very long and lustrous and the breed is found mainly in the north of England. The fleece is suited to combing due to its length and is used in specialty products due to its limited availability. The fineness varies from approximately 38–44 microns and length of 150–380 mm.

Suffolk

This UK breed is classified as short-wool and down. It is the most widely distributed breed of all the British breeds. This breed can be found throughout North America. Its dense fleece is suited to knitwear and any other application where a good bulk is required. It is a cross-breed of Norfolk and Southdown and has become a breed in its own right. The wool of the Norfolk was used in the original East Anglian cloths, which were made in the medieval times, which is where many of the cloths were produced at that time.

How to Market the Wool

If you are raising sheep of any breed for wool, you must have a marketing plan for the wool. While many clothes are made of synthetic fibers, there is still a need for the wool that sheep produce. Not as great a need as there was 500 or 5000 years ago, but it still exists.

Recently the need for wool is beginning to increase. Why? People all over the world have a demand for wool socks. Wool socks alone are increasing the need for wool production by a substantial amount. People are buying wool socks no matter the season because they have found these socks are especially nice for hiking. Wool socks sell anywhere from $20 to $30 a pair.

The fleece sheared off your sheep has to go through a lot of processing before it becomes one of those expensive pairs of socks, a baby blanket, a coat, or a new carpet. But getting that wool from sheep to sale isn't a one-size-fits-all situation.

You have to understand the various wool marketing channels, how to access them, and what qualities impact the pricing. The use of the fleece is imperative for all producers. While North America is an exporter of wool, with more than 70% of our domestically-produced commercial wool supply sold overseas, the demand for domestic wool is growing.

As stated above, the sock industry is booming with many brands touting their grown and sewn in the USA origins. Another game-changer was the recent introduction of the super-wash system, which rounds off the wool fibers by way of a specialized washing process, making them less itchy and less likely to shrink or stick together. Sheep producers both large and small can access the commercial wool market supply chain via various channels. Growers lacking in volume have some alternatives for pooling wool and selling into the commercial market. Pooling wool is where smaller farmers combine their wool to sell together as a larger community on the market. This sometimes will help a smaller farmer make more money.

Selling to Local Weavers

Some farmers sell to local weavers. If your operation is small, this may be the way to go. Contact the weavers who need sheep wool and ask how they need to have it made ready for them to buy. Often weavers will stop by farms that have sheep flocks and inquire about the wool.

You will get the lowest price for the wool if you sell it as a raw fleece, but that also represents the least amount of work. You need to skirt the fleece before selling, which means you pull off obvious manure tags and large vegetable matter (VM). This should not take you more than about five or ten minutes for each fleece, assuming it's relatively clean. It is not a good idea to sell fleeces that have sticks or thorns in them or those that are caked with mud—especially online when the buyer can't see the fleece before purchase.

Raw fleeces from breeds on The Livestock Conservancy's Conservation Priority List averages $16.63 per pound. Most ranged in price from $10 to $25 per pound with outliers as low as $5 and as high as $40 per pound. With raw fleeces, huge price variations may be due to the condition of fleece, meaning that the cheaper fleeces are not as clean and will require more work than the more expensive fleeces. If that is the way you decide to go, please note your sheep are not going to support you with their fleece.

Washed Fleece

By spending a good afternoon or early morning, a few gallons of water, and some laundry detergent, you can increase the sale price of your wool. Selling washed fleeces is not common, so if you decide to do this, you won't have a lot of competition. Most people sell fleece raw, as roving, or as yarn. Washed fleece were priced at $35 to $64 per pound with one Cotswold priced as high as $96 because the curly locks were being advertised as ideal to make wigs or doll hair.

If you know how to do spinning and can make and dye your own yarn, you can pull down very good money for your wool. The average price per pound of yarn is $80. Even if you hire help, you stand to make a decent profit if you have a large flock.

Conclusion to Raising Sheep

Whatever you decide in regards to raising sheep, this endeavor should be something you enjoy. If you are raising sheep for the idea of making lots of money, you should probably keep your day job if you have never farmed before. But if farming is something you do and you've decided sheep is the way to go to add cash to your farm savings, you could be right. If this is your first flock of sheep, start out small and build up a few head of sheep at a time.

As you read in our chapters, there are quite a few steps that go into the seasonal care of each sheep. Between the lambing and vaccinations, the parasite problems that can happen, and many other things, a shepherd or farmer must be diligent in taking care of a flock who is trusting to him or her. You may have a guardian animal or two or even three. But the bulk of care and safety still weighs on the shepherd.

Looking back over this book you can observe how important the relationship is between the sheep and the caregiver. The sheep should be able to trust you while they live on your farm. They actually depend on you for the correct food, medicine, shelter, and clean, parasite-free grazing areas and food. Fencing lines should be checked every few days and so should water that is standing in fields if you have very hot weather.

If you are planning to let a younger person of the house show one of the sheep for 4H, make sure the animal they are showing is being trained by the 4H child and he or she is doing the bulk of the work (within reason) to raise the lamb. Usually the lamb is then auctioned off at the auction. The going market rate will typically determine the approximate price you will get on your sheep, but there is no guarantee. Whether you are raising sheep for fun or for profit, you will realize that sheep are smart, funny, and unique creatures.

Sheep are docile, gentle animals for the most part. They are fairly economical in that they do not need perfect pasture for grazing. Fencing is not terribly expensive and neither are the barn structures. Sheep are usually much less difficult to raise than most other livestock and the profit per animal can be just as much if you understand what you are doing.

As a hobby animal, sheep are wonderful to have around your farm or home. They make wonderful pets and companions. Sheep get along well with dogs and other livestock, so they are a good compliment on a hobby farm.

Whatever your reasons for having sheep or learning about sheep, we hope you have found this guide to be useful and full of solid information. This book has taken you from the many kinds of sheep up through how to market the sheep you raise. Not only did you learn about the process of lambing but you learned about the vaccinations and the many parasites you should watch for in your sheep.

Raising sheep is sometimes difficult, but always rewarding in the sense that if you spend your time with the sheep, they will know you and will be easier to care for under critical times such as shearing, giving birth, or treating for medical problems. Use this guide as a reference and make sure to use other shepherds and farmers as your best resources. Have fun raising sheep and good luck!

Sources Used

Umberger, Steven H. 2009. Virginia Cooperative Extension Materials for Public use and Reprint. *Lambing Management.* Retrieved on March 30 from: https://resources.ext.vt.edu/ **

Staff, Penn State Extension. 2019. *So You Want to Raise Sheep or Goats?* Retrieved on 04/15/2021 from: https://extension.psu.edu/so-you-want-to-raise-sheep-or-goats

Queck-Matzie, Terri. Successful Farming. 2020. *New Ideas for the Sheep industry.* Retrieved on 4/01/2021 from https://www.agriculture.com/livestock/new-ideas-for-the-sheep-industry

Timber Creek Farm. 2019. *Sheep Care on Small Farms and Homesteads.* Retrieved on 04/01/2021 from: https://timbercreekfarmer.com/raising-sheep-without-grazing-pastures/

Weaver, Sue. 2019. *The Skinny on Keeping Sheep As Pets.* Retrieved on 04/01/2021 From: https://www.storey.com/article/skinny-on-keeping-sheep-as-pets/

Staff. Oregon State University. 2013. *Internal Parasites in Sheep and Goats.* Retrieved on 04/01/2021 from: https://agsci.oregonstate.edu/sites/agscid7/files/em9055.pdf

**Virginia Cooperative Extension materials are available for public use, reprint, or citation without further permission, provided the use includes credit to the author and to Virginia Cooperative Extension, Virginia Tech, and Virginia State University. Scott P. Greiner, Extension animal scientist, Virginia Tech Publication originally written by Steven H. Umberger, Extension Animal Scientist.

Edwin J. Jones, Director, Virginia Cooperative Extension, Virginia Tech, Blacksburg; M. Ray McKinnie, Administrator, 1890 Extension Program, Virginia State University, Petersburg. May 2009.

Endnotes

i http://www.corriedale.org.au/Page.asp?_=2013%20Bendigo%20ASBA

ii https://afs.ca.uky.edu/livestock/sheep/breeds/delaine-merino

iiiiii https://www.chegg.com/flashcards/sheep-breeds-c167d8cf-3285-48c8-8a5b-7683de5dccd6/deck

iv https://www.triplelfinnsheep.com/ancestors

v http://www.floridacrackersheep.com/

vi https://katahdinsheep.com/about.html

vii https://www.chegg.com/flashcards/sheep-breeds-c167d8cf-3285-48c8-8a5b-7683de5dccd6/deck

viii https://www.puddleduckfarm.net/navajochurrosheep.html

ix https://ixchelbunny.blogspot.com/2015/01/magical-hungarian-racka-adventures.html

x https://www.britannica.com/animal/Rambouillet-breed-of-sheep

xi https://livestockconservancy.org/index.php/heritage/internal/southdown

xii https://www.livestockoftheworld.com/Sheep/Breeds.asp?BreedLookupID=3003&Screenwidth=1200

xiii https://www.countryfile.com/wildlife/mammals/native-british-sheep-breeds-and-how-to-recognise-them/

xiv https://wiltshirehornsheepstud.com/

xv https://www.iamcountryside.com/fences-sheds-barns/a-diy-mobile-sheep-shelter/

xvi http://www.sheep101.info/201/housing.html

xvii https://www.iamcountryside.com/fences-sheds-barns/a-diy-mobile-sheep-shelter/

xviii https://www.pinterest.com/pin/5136987063773044/

xix https://uglydogsfarm.blogspot.com/2011/01/sheep-feeder-design.html

xx https://profence.org/animal-fencing/sheep-fence/

xxi https://www.gallagher.eu/en_gb/electric-fence-sheep

xxii https://wbaryfence.com/Electric-and-Wire-Fence/

xxiii https://www.stopgapfencing.co.uk/sheep_goat_fencing.html

xxiv http://www.crookwellvet.com.au/AnimalCare/Sheep/FlyStrike.aspx

xxv https://www.valleyvet.com/ct_detail.html?pgguid=f5e90a4e-571c-41da-9843-2d7989974f29

xxvi https://farmingwithcarnivoresnetwork.com/guardian-llamas/

xxvii https://www.thesprucepets.com/types-of-working-dogs-1118684

xxviii https://www.wideopenpets.com/everything-you-need-to-know-about-the-great-pyrenees-dog/sheeps-and-pyrenean-mountain-dog-known-as-the-great-pyrenees-in/

xxix https://www.agric.wa.gov.au/livestock-parasites/barbers-pole-worm-sheep

www.ingramcontent.com/pod-product-compliance
Lightning Source LLC
Chambersburg PA
CBHW081824200326
41597CB00023B/4371